International Conference on Nanomaterial, Semiconductor and Composite Materials (ICNSCM 2016)

MATEC Web of Conferences Volume 65 (2016)

Singapore
18-19 June 2016

Editors:

R. K. Agarwal

H. Abd. Rahman

ISBN: 978-1-5108-2602-1

Printed from e-media with permission by:

Curran Associates, Inc.
57 Morehouse Lane
Red Hook, NY 12571

proceedings.com

Some format issues inherent in the e-media version may also appear in this print version.

Printed by Curran Associates, Inc. (2016)

For additional information, please contact EDP Sciences – Web of Conferences
at the address below.

EDP Sciences – Web of Conferences
17, Avenue du Hoggar
Parc d'Activité de Courtabœuf
BP 112
F-91944 Les Ulis Cedex A
France

Phone: +33 (0) 1 69 18 75 75
Fax: +33 (0) 1 69 28 84 91

contact-edps@webofconferences.org

Additional copies of this publication are available from:

Curran Associates, Inc.
57 Morehouse Lane
Red Hook, NY 12571 USA
Phone: 845-758-0400
Fax: 845-758-2633
Email: curran@proceedings.com
Web: www.proceedings.com

TABLE OF CONTENTS

ICNSCM2016 Committee

Conference Chair

Jenny Ji, BOSI Education & Consultancy, China

Scientific Editor

Prof. Ramesh K. Agarwal, William Palm Professor of Engineering, Washington University in St. Louis, USA

Assoc. Prof. Dr. HAMIMAH ABD.RAHMAN, Universiti Tun Hussein Onn Malaysia, Malaysia

Program Committee

Ana Isabel Almerich Chulia, UNIVERSITAT POLIT CNICA DE VALèNCIA

Fahrettin Yakuphanoglu, Firat University

Yung Ting, Chung Yuan Christian University

Wenjie Chen, Singapore Institute of Manufacturing Technology

Xiufeng Han, Institute of Physics, Chinese Academy of Sciences (CAS)

Ching Yern Chee, University of Malaya

Ka-Veng YUEN, University of Macau (UM)

Niyazi OZDEMIR, Firat University

AZZURA BINTI ISMAIL, Universiti Tun Hussein Onn Malaysia, Malaysia

Prof. & Dr. Rehana Sharif, University of Engineering and Technology Lahore, Pakistan

Achanai Buasri, Silpakorn University, Thailand

Kunwar Singh Bartwal, Raja Ramanna Centre for Advanced Technology (RRCAT)

Sponsor

BOSI Education & Consultancy Co, Ltd

Preface

Dear Distinguished Authors and Guests,

The Organizing Committee warmly welcomes you to 2016 The International Conference on Nanomaterial, Semiconductor and Composite Materials (ICNSCM2016), held on June 18-19, 2016 in Singapore.

The aim of ICNSCM2016 is to present the latest research and results of scientists (professors, students, PhD Students, engineers, and post-doc scientist) related to Nanomaterial, Semiconductor and Composite Materials. This conference provides opportunities for the different areas delegates to exchange new ideas and application experiences face to face, to establish business or research relations and to find global partners for future collaboration.

The conference program is extremely rich, featuring high-impact presentation. We hope that the conference results constituted significant contribution to the knowledge in these up to date scientific field.

The proceeding records the fully refereed papers presented at the conference. The main conference themes and tracks are mechanical structures and smart materials and other correlation technique. Hopefully, all participants and other interested readers benefit scientifically from the proceedings and also find it stimulating in the process.

Finally we wish all the authors and attendees of ICNSCM2016 a unique, rewarding and enjoyable memory at ICNSCM2016 in Singapore. We look forward to your participation in the 2nd ICNSCM in 2017.

With our warmest regards,

Editor of ICNSCM2016

DOI: 10.1051/matecconf/20166501001

Self-Healing Concept for Damaged Composite Structure of Automobile Bonnet

Hyunbum Park[1,a]

[1]*Department of Defense Science &Technology - Aeronautics, Howon University, 54058, 64 Howondae 3gil, Impi, Gunsan, Rep. of Korea*

Abstract. In this work, a structural design and analysis on automobile bonnet using natural flax fiber composite is performed. Through the structural analyses using commercial FEM software, it is confirmed that the designed automobile bonnet using natural composite is acceptable for structural safety. And also, the damage is a critical problem in composites during their service in structural applications. Therefore, study on self-healing concept of bonnet structure was performed.

1 Introduction

For last several years, the interest to use natural fibers has been increased due to environmental problems of man made fibers in the composite materials. Natural fibers have some advantages such as cost effective, recycling and low energy consumption due to obtainable from nature rather than man made fibers by using high energy consumption.

In this work, a structural design on automobile bonnet using natural flax fiber composite is performed. The structural design results of flax/epoxy composite blade are compared with the design results of metal structure. In order to evaluate the structural design results of the composite bonnet, the structural analysis was performed by the finite element method. And also, the damage is a critical problem in composites during their service in structural applications. Therefore, study on self-healing concept of bonnet structure was performed.

2 Investigation on Natural Fiber

The growth of use of natural composite is indicative of their wider application due to the favorable mechanical performance of natural fibers. Among them, flax, hemp, jute and sisal are most popular composite fibers in natural composites [1~3]. Through comparison of mechanical strength and stiffness, it is found that the flax fiber has the best properties. The vinyl ester is selected due to cheap price even though somewhat low mechanical properties compared to epoxy. The tension test of specimen is shown as Fig. 1 and the stress-strain curves are shown as Fig. 2.

Figure 1. Manufactured flax/vinyl ester panel.

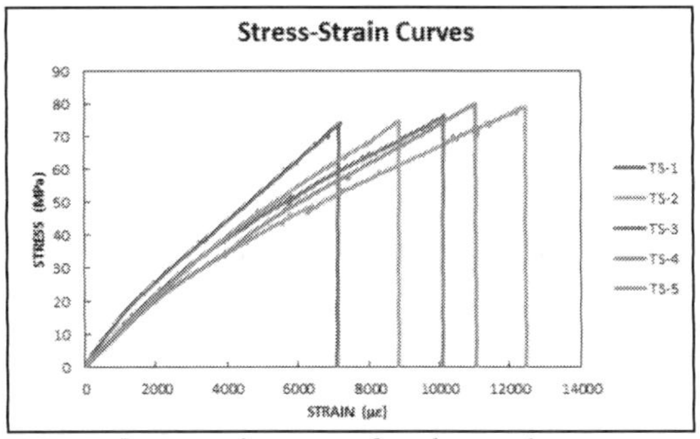

Figure 2. Stress-strain curves of tension specimen.

3 Structural Design and Manufacturing

After investigation on mechanical properties of flax/vinyl ester composite, the design of eco-friendly structure using flax/vinyl ester was performed. The selected target structure is bonnet panel for automobile. In order to evaluate the structural design results of bonnet, the

[a] Corresponding author: swordship@daum.net

structural analysis was performed by the finite element method. Through the structural analyses, it is confirmed that the designed bonnet using natural flax composite is acceptable for structural safety and stability. Figure 3 shows structural configuration of automobile bonnet.

The resin flow analysis of VARTM manufacturing method was performed after structural design and analysis. According to flow analysis results, the prototype bonnet was manufactured. Figure 4 shows the process of resin flow analysis. Figure 5 shows 3-D modeling of mould for prototype manufacturing.

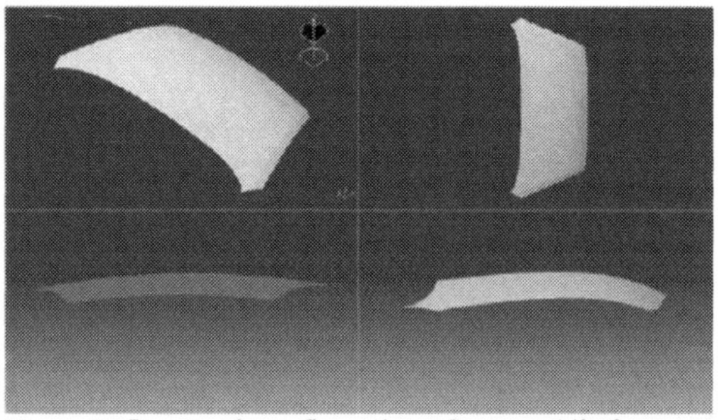

Figure 3. Structural configuration of automobile bonnet panel.

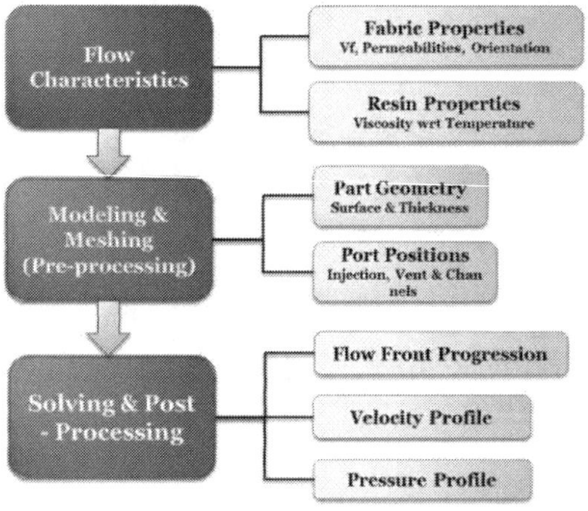

Figure 4. The process of resin flow analysis.

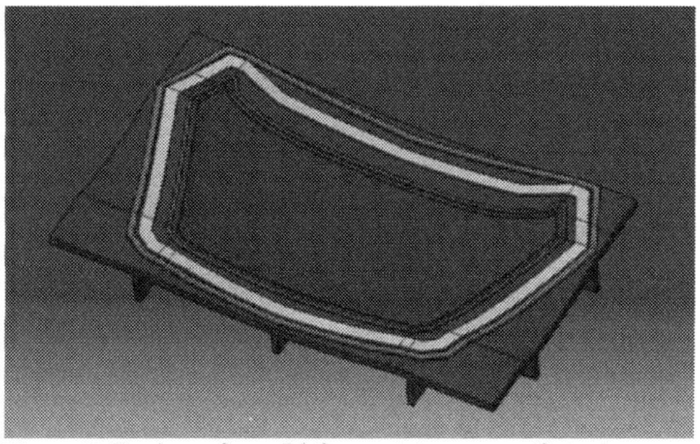

Figure 5. Design of mould for prototype manufacturing.

4 Investigation on Self-Healing of Composite Structure

The cracks in the structure are a critical problem in composite materials during their service in structural applications. The development of cracks brings about serious failure of the structures and then reduces their lifetimes. Therefore, early diagnosis of damage becomes necessary for removing the catastrophic failure. Recently, Self-healing composite structures have attracted increasing research interests. This study was performed investigation on self-healing of composite structure. Figure 6 shows the concept of self-healing epoxy based composite with epoxy loaded microcapsules. The self-healing concept using embedded microcapsules is shown in Fig. 6. A microencapsulated liquid monomer healing agent is embedded in a structural polymer matrix containing a catalyst capable of polymerizing the healing agent. When the material is damaged cracks occur, rupturing the microcapsules and releasing the healing agent into the crack plane through capillary action. The healing agent contacts the catalyst, triggering polymerization that bonds the crack faces closed[4, 5]. The target composite blade structure will be applied with the self-healing concept.

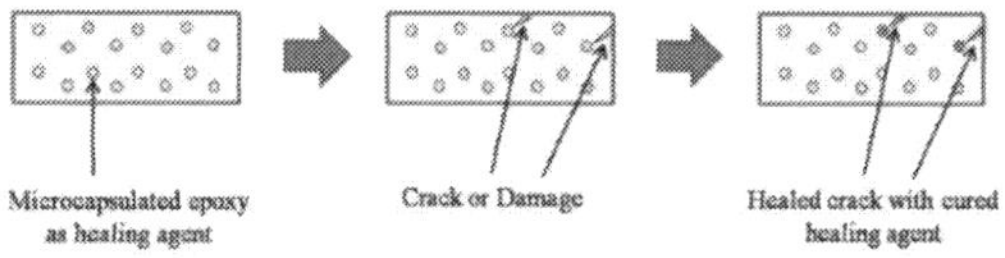

Figure 6. The concept of self-healing epoxy based composite with epoxy loaded microcapsules

5 Conclusions

In this study, an investigation on mechanical properties of flax/vinyl ester natural fiber composite is performed as a precedent study on the design of eco-friendly structure using flax/vinyl ester composite. The Vacuum Assisted Resin Transfer Molding (VARTM) manufacturing method is adopted for manufacturing the flax fiber composite specimen. The mechanical properties of the manufactured flax composites specimens are compared with flax composite data cited from some references. The experimental data show that the flax/vinyl ester composites using the proposed VARTM manufacturing method have much better mechanical properties than the reference test results. Based on this, structural design of automobile bonnet was performed using flax/vinyl ester. After structural design and analysis, the resin flow analysis of VARTM manufacturing method was performed. Through the structural analyses, it is confirmed that the designed bonnet is acceptable for structural safety and stability. Finally, study on self-healing concept of bonnet structure was performed.

Acknowledgments

This research was supported by Basic Science Research Program through the National Research Foundation of Korea (NRF) funded by the Ministry of Education (No. 2014R1A1A2054842). This study was supported by Howon University research funds.

References

1. L. Q. N. Tran, C. Fuentes, C. Dupont-Gillain, A. Van Vuure, I. Verpoest, Coir Fibre Composite: From Fibre Properties to Interfacial Adhesion and Mechanical Properties of Composites, Proceeding of ECCM, (2012)
2. D. Puglia, A. Iannoni, J. M. Kenny, C. Santulli, F. Sarasini, T. Valente, Thermal and Mechanical Behaviour of Phormium Tenax Reinforced Polypropylene Composite, Proceeding of ECCM, (2012)
3. Yanjun Xie, Callum A. S. Hill, Zefang Xiao, Holger Militz, Carsten Mai., Silane coupling agents used for natural fiber/polymer composites: a review, Composites Part A: Applied Science and Manufacturing, **41** (2010)
4. Y. C. Yuan, T. Yin, M. Z. Rong, M. Q. Zhang, Self healing in polymers and polymer composites. Concepts, realization and outlook: A review, eXPRESS Polymer Letters, **2** (2008)
5. M. R. Kessler, Self-healing: a new paradigm in materials design, Proc. IMechE Part G: J. Aerospace Engineering, **221** (2007)

A Study on Structural Design and Analysis of Blade and Tower for Small Wind Turbine System

Hyunbum Park[1,a]

[1]*Department of Defense Science &Technology - Aeronautics, Howon University, 54058, 64 Howondae 3gil, Impi, Gunsan, Rep. of Korea*

Abstract. This paper performed aerodynamic, structural design and analysis of the horizontal axis blade of low noise 1kW class small wind turbine system that can be operated at homes and designed to display excellent performances in low wind velocity regions such as Korea. In terms of the material applied, glass/epoxy that is both economical and excellent in its performance was applied. Structural test was conducted to verify whether the actual movement of the blade designed is consistent with the theoretical analysis result. In addition, structural safety review was performed through the structural analysis of a tower for installing the wind turbine system developed.

1 Introduction

Although wind power energy has been continuously used since ancient times, wind turbine system is being researched and developed in various ways recently in the midst of a current situation where studies are being actively conducted on alternative energy development resulting from the depletion of fossil fuel. Currently, a significant amount of power supply is being substituted through wind power generation centering on Europe and it is said that over 10% is being covered in regions such as Denmark and Northern Germany [1]. In terms of the research trend of wind power generation, wind turbine systems are being developed in various scales from a small system to a large power generation system. Since a small wind turbine system can be operated independently in a small scale, related studies are being actively conducted in recent [2]. In addition, the use of composite materials has led to the result of a significant improvement of structural strength and fatigue life. However, majority of the blade of small wind turbine system has been developed in advanced countries and the rated wind speed of imported blades is not suitable for the domestic climate condition. Accordingly, there is a need to develop wind turbine system suitable for the domestic climate condition.

This paper performed aerodynamic, structural design and analysis of the horizontal axis blade of low noise 1kW class small wind turbine system that can be operated at homes and designed to display excellent performances in low wind velocity regions such as Korea. In terms of the material applied, glass/epoxy that is both economical and excellent in its performance was applied. Structural test was conducted to verify whether the actual movement of the blade designed is consistent with the theoretical analysis result. In addition, structural safety review was performed through the structural analysis of a tower for installing the wind turbine system developed.

2 Blade Design

For the purpose of designing wind turbine system, it is necessary to first specify the system specification. As for the specification of the wind turbine system in this paper, it is horizontal axis wind turbine system with rate output of 1kW, rated wind velocity of 12.8m/s and applied airfoil NACA 632-615. The blade diameter is 2.54m and angle of torsion is 22.6° . The turbine system to be applied is a direct-driven AFPM (axial flux permanent magnet) turbine system that can be applied as a high-efficiency energy conversion device even at low wind velocity.

Structural design was performed after blade aerodynamic design to evaluate the structural safety of blade through structural analysis and fatigue life analysis. Structural load for structural design includes the aerodynamic load applied to blade and centrifugal load as main load.

Based on the structural design load, skin-spar-foam sandwich structure was adopted as the structural form of blade for the design. In the blade structural design form, skin and spar are stacked in the blade longitudinal with foam applied inside the blade, and it was designed to reduce the number of layers through longitudinal from the root part of blade for weight reduction by calculating section-specific load.

For the structural analysis conducted to evaluate the structural safety of structural design result, finite element commercial code was used, and Tsai-Wu failure criterion [3] was applied as the failure criterion for the safety review. The result of linear static analysis for each load

[a] Corresponding author: swordship@daum.net

condition confirmed that it has been designed with a safe structure ensuring sufficient safety factor. Natural frequency was analyzed to examine any resonance of blade and the result showed that resonance did not occur around operational number of rotations. The result of buckling stability evaluation also confirmed that it is a structure ensuring a sufficient level stability during operation with over 4 times the load. The maximum fatigue load calculated based on Spera's empirical formula for fatigue life analysis was 576.6Nm and the analysis result of stress on fatigue load showed maximum compressive stress of 35.4MPa and maximum tensile stress of 28.0MPa, thereby confirming that a sufficient level of safety rate has been ensured regarding allowable fatigue strength (144.7MPa). Accordingly, the blade design result confirmed that it is a structure that ensured required fatigue life of 20 years.

The structural analysis result was compared with the final structural test result for verification and they were consistent. The linear static analysis result and fatigue life analysis result were specified respectively in Fig. 1.

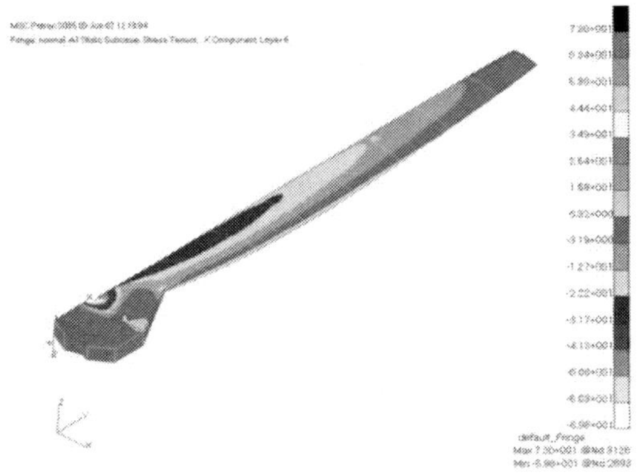

Figure 1. Stress analysis result of designed blade.

3 Tower Design

Structural analysis was performed for the tower that supports the blade system of this study to evaluate its structural safety [4]. The structural analysis of tower was performed by first determining load by considering each load condition and analyzing stress and displacement analysis after verifying the validity of load. As for load condition according to each situation, load of tower can be considered according to wind velocity changes. In this paper, two types of cases were considered with the load placed on tower during normal operation (load case 1) and the load placed on tower when it stopped during a storm (load case 2). Specific loads that need to be considered during tower structural analysis are load from thrust generated as blade rotates, and distributed load placed on tower from wind and load placed from nacelle & blade and tower weight they can be considered as non uniform load with concentrated load placed in the upper part of tower and distributed load placed on the entire tower. Fig. 2 shows the entire shape of tower. The structural analysis results showed that stress placed on tower during normal operation is 23.4 MPa and the stress on tower when it stopped during a storm is 154 MPa, thereby confirming that it is a structure that ensured a sufficient level of operational safety. Fig. 3 and 4 shows the results of structural analysis for each load condition and Fig. 6 is a view of blade and wind turbine system manufactured in this paper installed in a tower.

Figure 2. Structural configuration of automobile bonnet panel.

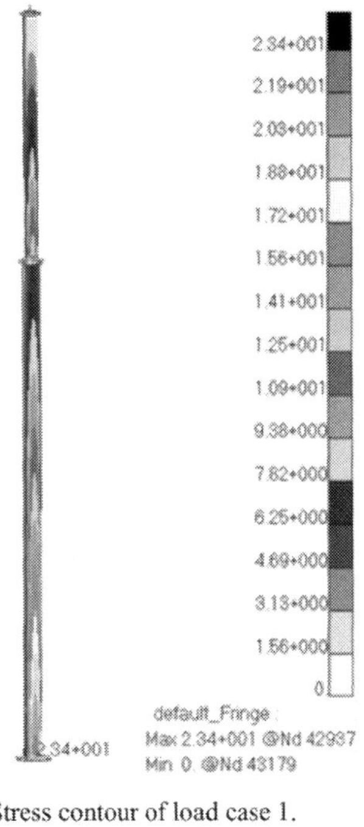

Figure 3. Stress contour of load case 1.

DOI: 10.1051/matecconf/20166501002

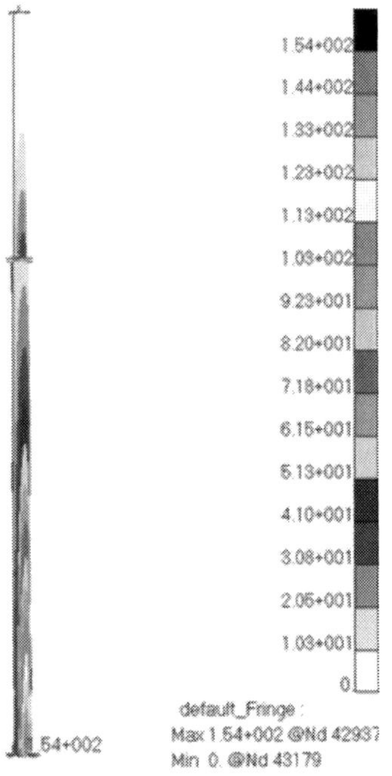

Figure 4. Stress contour of load case 2.

4 Conclusions

This paper performed aerodynamic, structural design and analysis of the horizontal axis blade of low noise 1kW class small wind turbine system that can be operated at homes and suitable for domestic regions. The result of reviewing its structural safety based on structural analysis showed that it is a structure that ensured a sufficient level of safety. In addition, the result of fatigue life analysis showed that it sufficiently satisfied the required life of 20 years. Based on the structural analysis of a tower supporting the blade system, it was confirmed that the tower structural design is a structure that ensured a sufficient level of safety.

Acknowledgments

This research was supported by Basic Science Research Program through the National Research Foundation of Korea (NRF) funded by the Ministry of Education (No. 2014R1A1A2054842). This study was supported by Howon University research funds.

References

1. D. LE Gourieres, *Wind power plants*, (Pergamon Press, 1982)
2. Changduk Kong, Johyuk Bang, Kyungwon Oh, Structural Design of 1kW class small wind turbine blade, Proceeding of Fall Conference on The Korean Society of Aeronautical & Space Science, (2003)
3. Fu-Kuo Chang, Richard A. Scott, George S. Springer, Strength of Mechanically Fastened Composite Joints, Journal of Composite Materials, **16** (1982)
4. Changduk Kong, Structural Design of Medium Scale Composite Wind Turbine Blade, Proceeding of 13th International Conference on Composite Materials, (2001).

Comparative Study of Gearbox Fault Diagnosis by Vibration Measurements

Dennis Hartono[1,a], Dunant Halim[1], Achmad Widodo[2], and Gethin Wyn Roberts[3]

[1]*The University of Nottingham Ningbo China, Mechanical, Materials and Manufacturing Engineering Department, 315100 Ningbo, China*
[2]*Universitas Diponegoro, Mechanical Engineering Department, 50275 Semarang, Indonesia*
[3]*The University of Nottingham Ningbo China, Civil Engineering Department, 315100 Ningbo, China*

Abstract. Vibration analysis has been demonstrated to be one of the best tools to detect faults in a gearbox by providing abundant information about the operating condition of a gearbox. However, a gearbox generates complex vibration signals, which makes it difficult to diagnose when a fault occurs. There are several fault diagnosis methods that can be utilized to analyze the underlying signals. The time-frequency method has been used and showed some promising results. On the other hand, it also has its drawback when it is applied to a complex mechanical system such as gearboxes. This paper thus attempts to examine the effectiveness of several diagnosis methods to detect faults in a gearbox from vibration measurements. The results show that the cepstrum method can provide a more accurate indication of a faulty gearbox compared to other diagnosis methods.

1 Introduction

The gearbox as one of the main parts of rotating machinery has been used in many engineering applications, requiring speed and torque conversions. The failure of the gear mechanism affects the entire operation of the machinery, which can cause significant loss in industry. Vibration measurement is considered as the most common method to detect failure in gears. A number of diagnosis methods have been proposed to analyze vibration signals from a defective gearbox. The frequency domain approach such as the Fast Fourier Transform (FFT) can offer a different perspective to diagnose faults in the gearbox. Theoretically it works well for stationary-signals, but not for non-stationary (transient) signals. On the other hand, the time-frequency domain, *e.g* Continuous Wavelet Transform (CWT), can provide a reliable diagnosis for gearboxes operating under a non-stationary condition (the run-up, run-down and transient processes) [1,2]. The simultaneous representation in time and frequency domains in a single chart can also help to provide a more detailed diagnosis of occurring faults. Unfortunately none of the methods in the time-frequency domain can provide a complete solution to indicate faults and sometimes the results can also be misleading [3].

When the faulty gear is in contact with a normal gear, a transient impulse vibration will be generated. It is well known that when the gear is faulty, sidebands can be observed around its gearmesh frequency and its harmonics. The spacing of the sidebands indicates the location of the faulty gear [4]. However it is not an easy task to visualize such a sideband by using a conventional frequency domain method such as the Fourier Transform. This work thus compares several methods to diagnose a faulty gearbox and suggests the most effective way to diagnose the fault.

2 Fundamental Theory

2.1.1 The Continuous Wavelet Transform

The CWT is able to overcome the limitation from the FFT which only can offer global feature of the underlying signal x(t). The CWT moves data from a space to a scale domain. One of the main advantage of using CWT is its ability to provide high frequency resolution low frequency region and high time resolution in high frequency region which is very useful for gearbox fault diagnosis. The CWT of a signal x(t) is defined as a convolution integral of x(t) with dilated and scaled versions of a mother wavelet function Ψ(t) [5]:

$$CWTx(a,b) = {}^{1}\!/_{\sqrt{a}}\int_{-\infty}^{\infty} x(t)\Psi^{*}(\tfrac{t-b}{a})d\tau \quad (1)$$

Where a and b is the scale parameter and translation parameter respectively. Equation 1 indicates that CWT is similar to the Fourier Transform except that the CWT

[a] Corresponding author: dennis.hartono@nottingham.edu.cn

uses the family of wavelets function instead of using sine and cosine function as the basis functions.

2.1.2 The Cepstrum

The Cepstrum was initially used to detect echo delay times in seismic signals [6]. As mentioned in the Introduction, localized fault in the gearbox will give rise to sidebands in the vicinity of the gearmesh frequency and its harmonics. Therefore sidebands spacing contains useful information since it is related to the faulty source. The Cepstrum possess the ability to detect sidebands without a thorough inspection in frequency domain. Randal mentioned that there are 3 main applications of cepstrum analysis in gear diagnostics [4], however author believe that in Condition Monitoring of a gearbox, the main application of the Cepstrum analysis is to detect the presence of such sidebands family and its harmonics [4, 7]. Initially The cepstrum was defined as the power spectrum of the logarithm of the power spectrum. However in the area of gearbox fault diagnosis, the Cepstrum is commonly defined as the inverse Fourier Transform of the logarithmic power spectrum as defined as equation two below. Randall [4] mentioned that it is partly because it is more logical to use the inverse transform between a function of frequency and a function of time, and partly because it is then reversible to the power spectrum.

$$C(\tau) = F^{-1}\{log[Sxx(f)]\} \quad (2)$$

Where Sxx(f) is the power spectrum. The indepcncent variable τ known as quefrency has the dimension of time(s).

3 Comparisons of Several Fault Diagnosis Methods

A single-stage spur gearbox was analyzed in this work. Three Accelerometers connected to the DAQ SpectraPed system were used to capture the vibration signal from the gearbox with the 5 kHz sampling frequency. The accelerometers were mounted on one of bearings. The fault investigated in this experiment was the 50% tooth removal gear, located at the output of the gearbox with the rotational frequency of 16.67 Hz. The goal of this experiment was to detect the particular output rotational frequency or its harmonics.

The vibration signals (in terms of the voltage output from accelerometers) from the healthy gearbox and faulty gearbox are presented in Figure 1 and Figure 2, respectively. There was a slight increase in vibration magnitude when a fault existed in the gearbox. An impulse-like response (spike) could occur when the faulty gear tooth make a contact with the other gear. However, it was difficult to identify the periodicity of this particular spike (which could correspond to the faulty gear tooth) by a direct visual inspection of the plot. It was not possible to obtain any information about location of the faulty gear tooth from the time-domain representation.

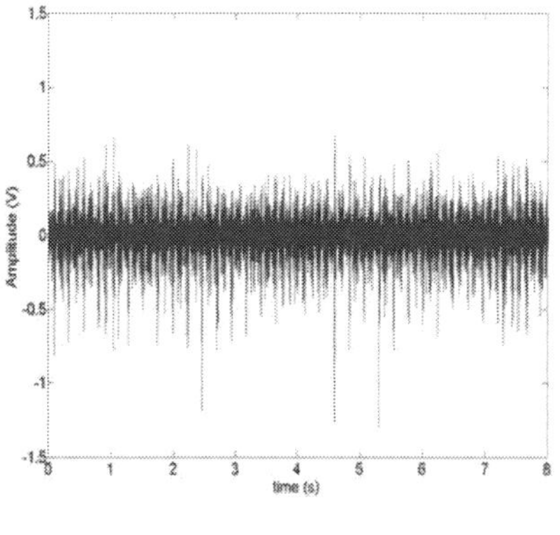

Figure 1. Vibration signal from a normal gearbox

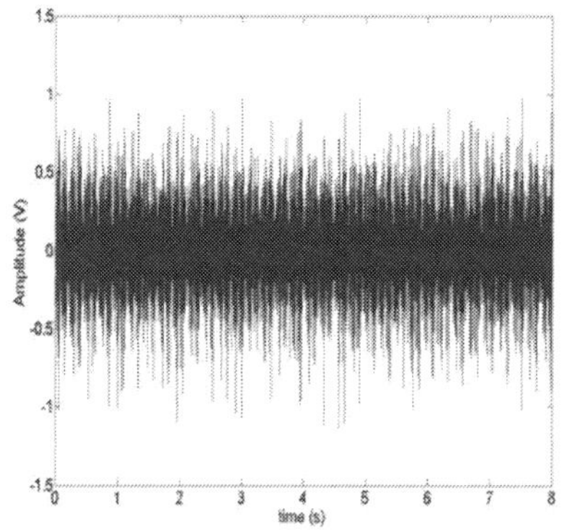

Figure 2. Vibration signal from a faultyl gearbox

The Fourier Transform offers a different representation of signals in frequency domain. It gives an indication of the faulty gearbox from the sideband spacing around the gearmesh frequency and its harmonics. The gearmesh frequency of the gearbox was 250 Hz. However it was difficult to identify the sideband spacing from the FFT due to the signal noise as shown in Figure 3. Moreover, some sidebands could be covered by the noise, which had higher amplitudes compared to those of the sidebands. There were too many spurious peaks that also appeared in the plot that made it even harder to analyse. The spectral leakage had also an inherent drawback that exacerbated the FFT plot. Although this could be mitigated by applying a windowing function, it could not be eliminated completely. Furthermore, it was not possible to know when such a frequency existed from The Fourier Transform. It is an inherent drawback from the FFT because by default FFT utilizes sine and cosine as a basis function which means it only can provide global feature of the underlying signal which means we will lose the local feature that is represented in time-domain. To obtain such information, another method that generates the time-frequency information will be needed.

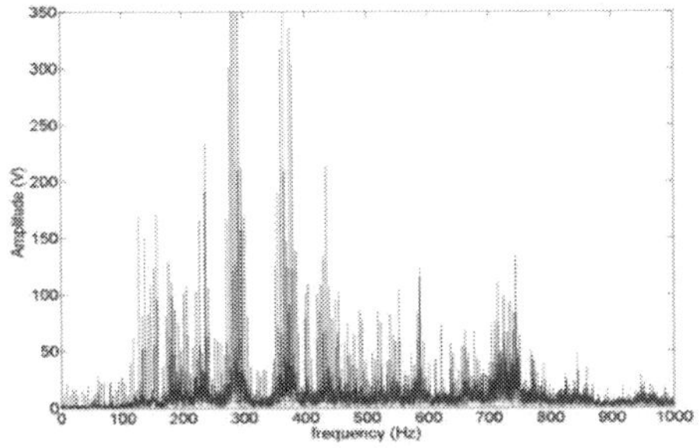

Figure 3. FFT of vibration signal for a faulty gearbox

To address the drawback of The Fourier Transform, The Continuous Wavelet Transform was investigated, for providing a time-scale (time-frequency) representation. Figure 4 shows the CWT of the normal gearbox. There was not any obvious periodicity from the figure. However it can be observed clearly that there were new generated contours for a faulty gearbox in Figure 5. The periodicity was equal to 0.24 s. This corresponded exactly to the fourth harmonic of the faulty gearbox rotation period, which was 4x(60/1000) s. This periodicity can be observed more easily than the sideband spacing shown in Figure 3. The CWT scales from 11 to 21 were related to a frequency range from 193.45 Hz to 369.31 Hz. From the CWT plot in Figure 5, a conclusion could be drawn that there were sidebands that appeared around the first gearmesh frequency. Although Figure 3 could also provide the same information (regarding the sideband spacing around the gearmesh Frequency), it was more difficult to identify due to the existence of noise that can mask the sidebands.

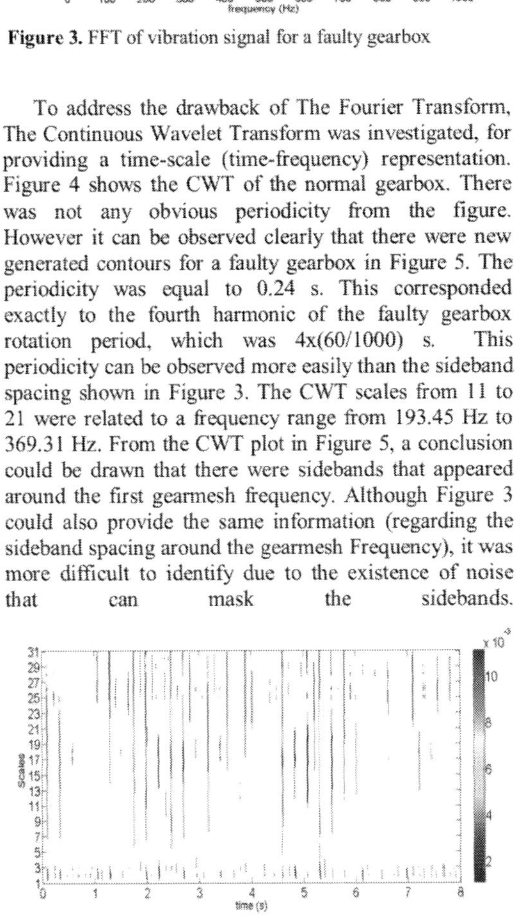

Figure 4. CWT of vibration signal for a a normal gearbox

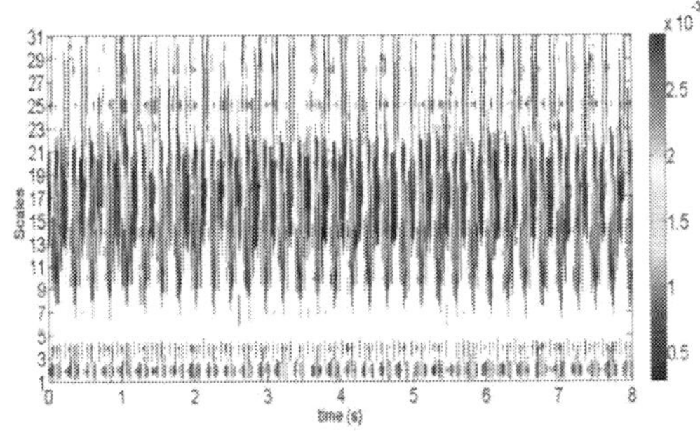

Figure 5. CWT of vibration signal for a a faulty gearbox

Finally, the cepstrum method measures the periodicity in the frequency domain (Fourier Transform)e.g. the sideband spacing around the gearmesh frequency and its harmonics[4]. The result from CWT was also supported by the cepstrum results in Fig. 6. It was obvious that there was a peak at 0.2375s, which corresponded exactly to the fourth harmonic of the faulty gear. Moreover, its harmonics could be observed at 0.4756s, 0.7131s, and 0.951s. The cepstrum method provided the most direct way in detecting a fault, compared to need of a visual inspection in the CWT method.

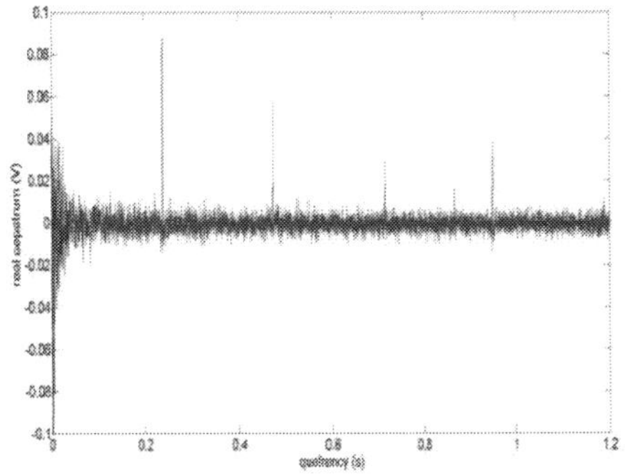

Figure 6. The cepstrum of vibration signal for a faulty gearbox

4 Summary

In this work, several diagnosis methods have been investigated to detect faults in a gearbox. The Fourier Transform was not informative enough to identify the faults in a gearbox, particularly due to the signal noise that masked the sidebands. The Continuous Wavelet transform, on the other hand, could provide a different perspective in time-frequency diagram. The pattern of the faulty gearbox could be directly observed from the contour of the diagram, indicating the corresponding harmonics of the faulty gear. Nevertheless, it required as visual inspection and the need to zoom in to the diagram to identify the fault indicator. The cepstrum method, however, could provide direct fault identification without the need of a thorough visual inspection of the CWT method.

Acknowledgement

The authors gratefully acknowledge the financial support provided by Ningbo Science and Technology Bureau-International Cooperation Programme (Project: 2012D10029), China.

References

1. H. Zheng, Z. Li, X. Chen, Mec Sys and Sig Proc **16**, 447-457 (2002)
2. R. Yan, R. X. Gao, X. Chen, Mech Sys and Sig Proc **96(a)**, 1-15 (2014)
3. Feng, Z, M. Liang, Chu. F, Mech Sys and Sig Proc **38(1)** 165-205 (2013)

4. R. B. Randall, *Vibration Based Condition Monitoring: Industrial, Aerospace and Automotive Applications* (2011)
5. N. Baydar and A. Ball, Mech Sys and Sig Proc **17(4)**, 787-804 (2003)
6. B. P. Bogert, M. J. R Healy, J. W. Tukey. *In Proocedings of the Symposium on Time Series Analysis*, 209-243 (1963)
7. G. Dalpiaz, A. Rivola and R. Rubini, Mech Sys and Sig Proc **14(3)**, 387-412 (2000)

Dynamic Model and Analysis of Asymmetric Telescopic Wing for Morphing Aircraft

Chen Lili[a], Guo Zheng, Wang Wenkai

Institute of Nearspace Technology, College of Aerospace Science and Engineering, National University of Defense Technology, Changsha , China

Abstract. Morphing aircraft has been the research hot topics of new concept aircrafts in aerospace engineering. Telescopic wing is an important morphing technology for morphing aircraft. This paper describes the dynamic equations and kinematic equations based on theorem of momentum and theorem of moment of momentum, which are available for all morphing aircrafts. Meanwhile,as simplified , dynamic equations for rectangular telescopic wing are presented. In order to avoid the complexity using aileron to generate rolling moment , an new idea that asymmetry of wings can generate roll moment is introduced. Finally, roll performance comparison of asymmetric wing and aileron deflection shows that asymmetric telescopic wing can provide the required roll control moment as aileron, and in some cases, telescopic wing has the superior roll performance.

1 Introduction

Morphing aircraft has been investigated extensively because of its superior flight performance[1, 2]. Morphing aircraft can adapt various complicated flight regimes.The Defense Advanced Research Projects Agency (DARPA) developed Morphing Aircraft Program in 1998. At the beginning of 21 century, NASA showed its morphing aicraft project. Ordinarily morphing aircraft can be classified three types:micro morphing、small morphing and large morphing. Micro morphing includes variable camber[3, 4] and twist[5], small morphing contains variable tip and variable horizontal tail, while variable sweep[6, 7],variable span[8] and folding wing belong to large morphing. This paper mainly concentrates on telescopic wing which is one of large morphing techniques.

Telescopic Wing is intended to change its wing span for various flight conditions to reduce drag. As aresult of increasing the wing span, the aspect ratio and wing area increase and the spanwise lift distribution decreases for the same lift[8].Telescopic wing has a long research history, many patents of invention have been declared[9], as shown in Fig.1, which is three patents about telescopic wing.

In this work, the dynamic and kinematic model are described to investigative performance of morphing aircraft. But, because the dynamic model involves some variable parameters with time, which leads to complexity of morphing aircraft. In order to obtain some basic characteristics, some hypotheses are put forward. Telescopic wing aircraft can change wing span and area during flight to realize better aerodynamic and control performance. Although much work has been done, technique of telescopic wing is not well developed.Some fundamental reseach work of telescopic wing was done focusing on dynamic model and analysis in this paper .

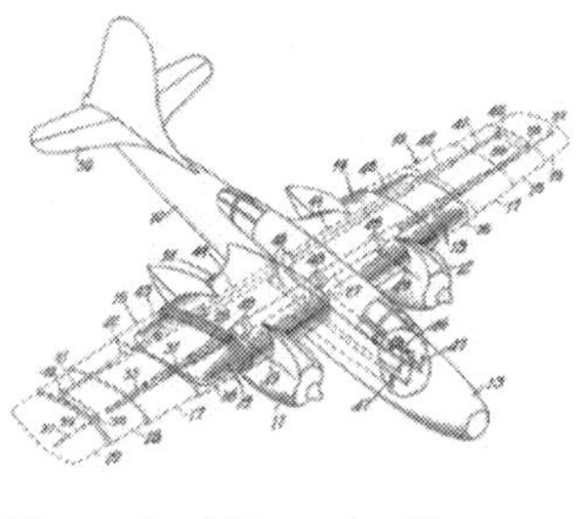

(a) Patent of variable erea by Gibson

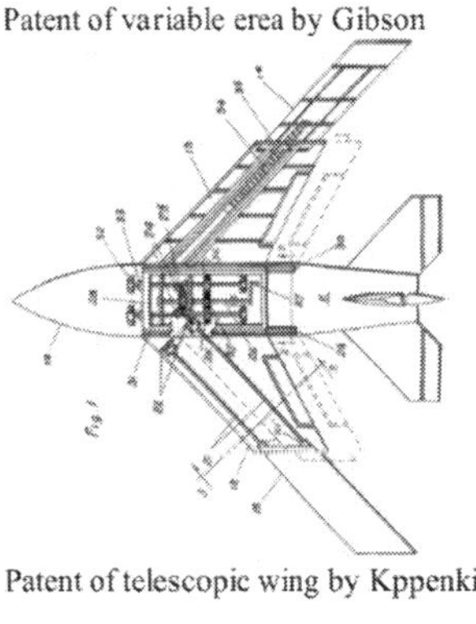

(b) Patent of telescopic wing by Kppenkin

[a] Chen Lili: doublely@foxmail.com

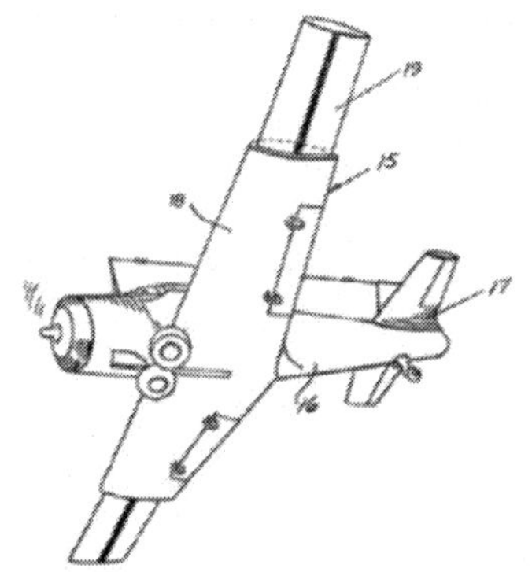

(c) Patent of variable erea by Murray

Figure.1 Patents of telescopic wings

2 Morphing aircraft model

2.1 Dynamic model

In this section the dynamic model of morphing aircraft is presented based on theorem of momentum and theorem of moment of momentum and also refers the symmetric wing dynamic model of ideal rigid-body, which is related to telescopic wing by Jin Dongping[10]. On the assumption that origin is located on symmetry position of fuselage, aircraft axes (or named as body axes)O-XYZ(as shown in Fig. 2) can be defined.In addition, inertial reference axes is defined as O_A-$X_AY_AZ_A$(as shown in Fig. 2) attached to ground (earch whirling ignored). The velocity and angular rate of morphing aircraft are V_A and ω_A relative to inertial reference axes,respectively. Usually,V_A and ω_A are projected on aircraft axes , marked as $V=(V_x,V_y,V_z),\omega=(\omega_x,\omega_y,\omega_z)$.

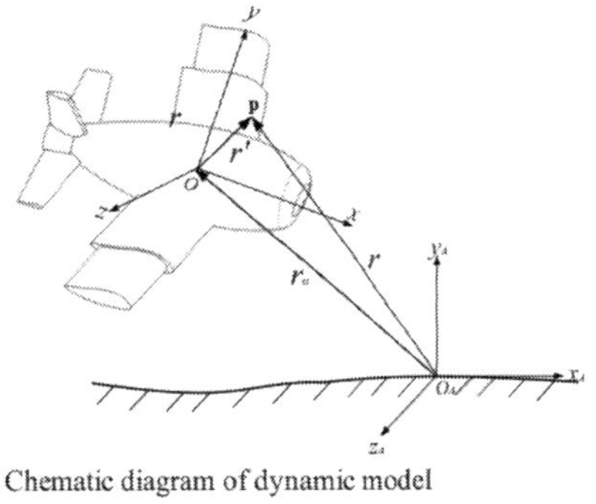

Figure 2 Chematic diagram of dynamic model

For any ***dm*** of body, its position vetor is $\boldsymbol{r}_i$ to inertial axes, the integral result can be expressed as:

$$p=\int_\Omega \dot{r}_i\,dm=\int_\Omega(\dot{r}_o+\dot{r}')dm \tag{1}$$

According to theorem of moment of mass, the latter part can be noted as:

$$S=\int_\Omega r'dm \tag{2}$$

Derivative expression is

$$\dot{S}=\overset{\circ}{S}_o+\omega\times S_o \tag{3}$$

Where $\dot{S}$ and $\overset{\circ}{S}_o$ are the derivative of instantaneous mass moment in inertial reference axes and aircraft axes, respectively. "$\circ$ "represents the derivative to time in the aircraft axes.

Duo to $\dot{r}_o=V$,thus

$$p=mV+\overset{\circ}{S}_o+\omega\times S_o \tag{4}$$

Where m is mass of aicraft (including fuselage and wings ,also other parts),the moment of momentum relative to origin of aircraft axes can be expressed:

$$\begin{aligned}L_o&=\int_\Omega r_i'\times v_i dm=\int_\Omega r_i'\times(V+\dot{r}_i')dm\\&=S_o\times V+\int_\Omega r_i'\times\dot{r}_i'\,dm\end{aligned} \tag{5}$$

Because aircaft axes has the rotation rate ω itself ,so

$$L_o=S_o\times V+\int_\Omega r_i'\times(\overset{\circ}{r}_i'+\omega\times r_i')dm \tag{6}$$

In the light of definition,

$$\begin{aligned}\int_\Omega r_i'\times(\omega\times r_i')dm&=\int_\Omega[\omega(r_i'\cdot r_i')\\&-r_i'(\omega\cdot r_i')]dm=I\cdot\omega\end{aligned} \tag{7}$$

Where I is the aircraft instantaneous inertia tensor relative to origin O,the expression is as follows:

$$I=\begin{bmatrix}I_{xx}&-I_{xy}&-I_{xz}\\-I_{yx}&I_{yy}&-I_{yz}\\-I_{zx}&-I_{zy}&I_{zz}\end{bmatrix} \tag{8}$$

So $$L_o=S_o\times V+I\cdot\omega+\int_\Omega(r_i'\times\overset{\circ}{r}_i')dm \tag{9}$$

Theorem of momentum: $$\dot{p}=F \tag{10}$$

To make it simple,the relative theorem of moment of momentum of any point can be demonstrated in the translation axes as in Fig.3:

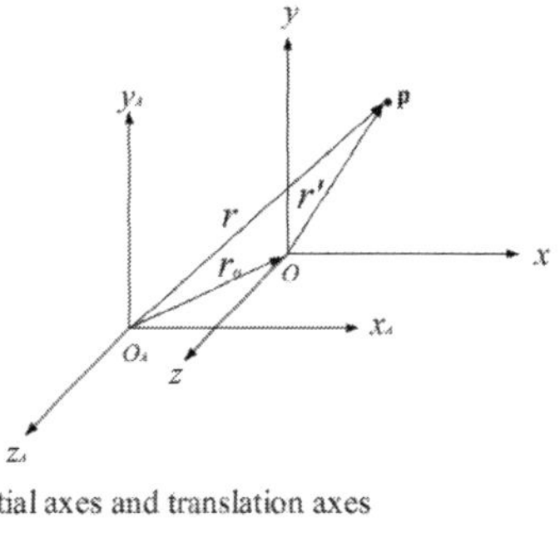

Figure 3 Inertial axes and translation axes

$$L_{Oi}=r_i'\times m_i v_i \tag{11}$$

Eq.(11) can be derivated ,applying Newton second law to the result:

$$\frac{dL_{Oi}}{dt}=\frac{dr_i'}{dt}\times m_i v_i+r_i'\times m_i\frac{dv_i}{dt}=\frac{dr_i'}{dt}\times m_i v_i+r_i'\times F_i \tag{12}$$

DOI: 10.1051/matecconf/20166501004

Due to

$$\frac{dr_i'}{dt}=\frac{dr_i}{dt}-\frac{dr_o}{dt}=v_i-v_o \tag{13}$$

$$r_i'\times F_i=M_{oi} \tag{14}$$

Substituting Eq.(13) and Eq.(14) into Eq.(12) and integral yields the theorem of moment of momentum of whole aircraft

$$\frac{dL_o}{dt}=\int_\Omega v_i dm\times V+M_o=p\times V+M_o \tag{15}$$

Substituting Eq.(4) and Eq.(9) into Eq.(15) and considering vetor cross product relationship:

$$\omega\times(S_o\times V)+V\times(\omega\times S_o)=S_o\times(\omega\times V) \tag{16}$$

Taking an inertial derivative of the vector with respect to time in whirling reference frame[11], morphing aircraft flight dynamic model afer simplification under body axes yields

$$\begin{cases} m(\dot{V}+\omega\times V)+\ddot{S}_o+\dot{\omega}\times S_o+2\omega\times\dot{S}_o \\ +\omega\times(\omega\times S_o)=F \\ I\cdot\dot{\omega}+\dot{I}\cdot\omega+\omega\times(I\cdot\omega)+S_o\times\dot{V}+S_o\times(\omega\times V) \\ +\int_\Omega\{r_i'\times\ddot{r}_i'+\omega\times(r_i'\times\dot{r}_i')\}dm=M_o \end{cases} \tag{17}$$

Where F is the force imposed on the aircraft under body axes, M_o is the force moment relative to the origin of body axes caused by external force. Eq.(17) shows that motion parameters of telescopic wing aircraft are tightly related with the time variable of moving wing,for instance, S_o、S_{Oi}、I are functions of time. To rigid aircraft,wings maintain the constant relative position,then the equations reduce as:

$$\begin{cases} m(\dot{V}+\omega\times V)+\dot{\omega}\times S_o+\omega\times(\omega\times S_o)=F \\ I\cdot\dot{\omega}+\omega\times(I\cdot\omega)+S_o\times\dot{V}+S_o\times(\omega\times V)=M_o \end{cases} \tag{18}$$

If ,furthermore, center of mass is selected as the origin of body axes, Eq.(18) can be simplified as:

$$\begin{cases} m(\dot{V}+\omega\times V)=F \\ I\cdot\dot{\omega}+\omega\times(I\cdot\omega)=M_o \end{cases} \tag{19}$$

This equation can be used as center of mass motion equation and around center of mass motion equation in noninertial system.

Generally, moving wings have a small part mass of the whole aircraft and that product of inertia is little relatively.So, it is proper that I_{xy}, I_{yz}, I_{zx} are equal to 0, scalar quantity format of dynamic equations in the body axes can obtain in the following:

$$\begin{cases} m(\dot{V}_x+\omega_y\cdot V_z-\omega_z\cdot V_y)+\ddot{S}_{ox}+(\dot{\omega}_y\cdot S_{Oz}-\dot{\omega}_z\cdot S_{Oy}) \\ +2(\omega_y\cdot\dot{S}_{Oz}-\omega_z\cdot\dot{S}_{Oy})+\omega_y\cdot(S_{oy}\omega_x-S_{ox}\omega_y) \\ +\omega_z\cdot(S_{oz}\omega_x-S_{ox}\omega_z)=F_x \\ m(\dot{V}_y+\omega_z\cdot V_x-\omega_x\cdot V_z)+\ddot{S}_{oy}+(\dot{\omega}_z\cdot S_{Ox}-\dot{\omega}_x\cdot S_{Oz}) \\ +2(\omega_z\cdot\dot{S}_{Ox}-\omega_x\cdot\dot{S}_{Oz})+\omega_z\cdot(S_{oz}\omega_y-S_{oy}\omega_z) \\ -\omega_x\cdot(S_{oy}\omega_x-S_{ox}\omega_y)=F_y \\ m(\dot{V}_z+\omega_x\cdot V_y-\omega_y\cdot V_x)+\ddot{S}_{oz}+(\dot{\omega}_x\cdot S_{Oy}-\dot{\omega}_y\cdot S_{Ox}) \\ +2(\omega_x\cdot\dot{S}_{Oy}-\omega_y\cdot\dot{S}_{Ox})-\omega_x\cdot(S_{oz}\omega_x-S_{ox}\omega_z) \\ -\omega_y\cdot(S_{oz}\omega_y-S_{oy}\omega_z)=F_z \end{cases} \tag{20}$$

$$\begin{cases} I_x\cdot\dot{\omega}_x+\dot{I}_x\cdot\omega_x+(\omega_y(I_z\cdot\omega_z)-\omega_z(I_y\cdot\omega_y))+(S_{Oy}\dot{V}_z \\ -S_{Oz}\dot{V}_y)-S_{Oy}(V_x\cdot\omega_y-V_y\cdot\omega_x)-S_{Oz}(V_x\cdot\omega_z-V_z\cdot\omega_x) \\ +\int_\Omega\{r_{iy}'\cdot\ddot{r}_{iz}'-r_{iz}'\cdot\ddot{r}_{iy}'+\omega_y(\dot{r}_{iy}'\cdot r_{ix}'-\dot{r}_{ix}'\cdot r_{iy}') \\ +\omega_z(\dot{r}_{iz}'\cdot r_{ix}'-\dot{r}_{ix}'\cdot r_{iz}')\}dm=M_{ox} \\ I_y\cdot\dot{\omega}_y+\dot{I}_y\cdot\omega_y+(\omega_z(I_x\cdot\omega_x)-\omega_x(I_z\cdot\omega_z))+(S_{Oz}\dot{V}_x \\ -S_{Ox}\dot{V}_z)+S_{Ox}(V_x\cdot\omega_y-V_y\cdot\omega_x)-S_{Oz}(V_y\cdot\omega_z-V_z\cdot\omega_y) \\ +\int_\Omega\{r_{iz}'\cdot\ddot{r}_{ix}'-r_{ix}'\cdot\ddot{r}_{iz}'+\omega_z(\dot{r}_{iz}'\cdot r_{iy}'-\dot{r}_{iy}'\cdot r_{iz}') \\ -\omega_x(\dot{r}_{iy}'\cdot r_{ix}'-\dot{r}_{ix}'\cdot r_{iy}')\}dm=M_{oy} \\ I_z\cdot\dot{\omega}_z+\dot{I}_z\cdot\omega_z+(\omega_x(I_y\cdot\omega_y)-\omega_y(I_x\cdot\omega_x))+(S_{Ox}\dot{V}_y \\ -S_{Oy}\dot{V}_x)+S_{Ox}(V_x\cdot\omega_z-V_z\cdot\omega_x)+S_{Oy}(V_y\cdot\omega_z-V_z\cdot\omega_y) \\ +\int_\Omega\{r_{ix}'\cdot\ddot{r}_{iy}'-r_{iy}'\cdot\ddot{r}_{ix}'-\omega_x(\dot{r}_{iz}'\cdot r_{ix}'-\dot{r}_{ix}'\cdot r_{iz}') \\ -\omega_y(\dot{r}_{iz}'\cdot r_{iy}'-\dot{r}_{iy}'\cdot r_{iz}')\}dm=M_{oz} \end{cases} \tag{21}$$

Obviously,the dynamic equation series is very complex and has 6-DOFs that couple with each other.It is difficult to solve. Whereas, the equation is available for all morphing aircrafts and can be applied to kinds of ridid or flexible morphing aircraft ,in the previous assumption.

2.2 Kinematic model

In order to obtain the movement trajectory of morphing aircraft in inertial axes,it is essential to bulid the relatinship of inertial axes and body axes.As usual,the relationship of the two axes is denoted by three Euler angles.Their definitions[12] (Fig.4) as follws:

Angle of pitch φ :the angle between the longitudinal axis of a flight vehicle ox and the horizontal surface $x_A oz_A$.

Angle of yaw ψ :the angle between the axis of ox_A and the longitudinal axis ox' projection on the horizontal surface $x_A oz_A$.

Angle of roll γ :the angle between the axis of oy

DOI: 10.1051/matecconf/20166501004

and vertical surface including the longitudinal axis ox of the fight vehicle.

Coordinate transformations[12] between inertial axes and body axes are shown as

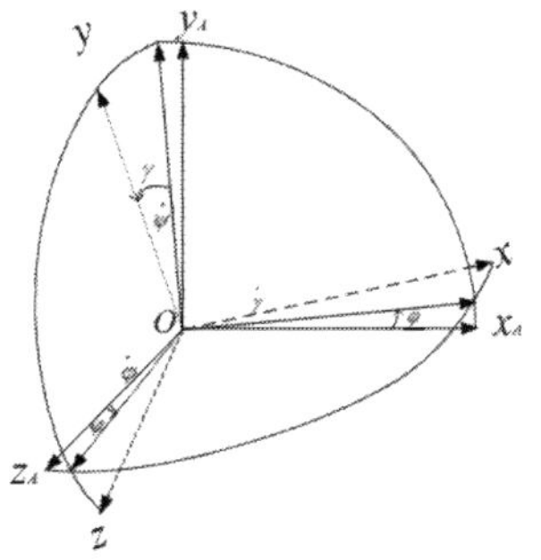

Figure 4 Relationship of Euler angles

$$B_A = \begin{bmatrix} \cos\varphi\cos\psi & \sin\varphi\cos\psi & -\sin\psi \\ \cos\varphi\sin\psi\sin\gamma - \sin\varphi\cos\gamma & \sin\varphi\sin\psi\sin\gamma + \cos\varphi\cos\gamma & \cos\psi\sin\gamma \\ \cos\varphi\sin\psi\cos\gamma + \sin\varphi\sin\gamma & \sin\varphi\sin\psi\cos\gamma - \cos\varphi\sin\gamma & \cos\psi\cos\gamma \end{bmatrix}$$

Then Kinematic equation in the inertial axes can be given by

$$\left[V_{Ax}, V_{Ay}, V_{Az}\right]^{\mathrm{T}} = A_B\left[V_x, V_y, V_z\right]^{\mathrm{T}}$$

Where $A_B = B_A^T$, the corresopnding scalar format is

$$\begin{cases} \dfrac{dx_A}{dt} = \cos\varphi\cos\psi V_x + (\cos\varphi\sin\psi\sin\gamma - \sin\varphi\cos\gamma)V_y \\ +(\cos\varphi\sin\psi\cos\gamma + \sin\varphi\sin\gamma)V_z \\ \dfrac{dy_A}{dt} = \sin\varphi\cos\psi V_x + (\sin\varphi\sin\psi\sin\gamma + \cos\varphi\cos\gamma)V_y \\ +(\sin\varphi\sin\psi\cos\gamma - \cos\varphi\sin\gamma)V_z \\ \dfrac{dz_A}{dt} = -\sin\psi V_x + (\cos\psi\sin\gamma)V_y + (\cos\psi\cos\gamma)V_z \end{cases} \tag{22}$$

The attitude of aircraft is described by three Euler angles φ、ψ and γ.The anglar equations of motion can be obtained throught finding the transformation relation of Euler angles[13].First, φ is formed by the angular rate $\dot{\varphi}$ around z_A axis, ψ is from the angular rate of new axis y', which is projected on body axes as $[0, \dot{\psi}\cos\gamma, -\dot{\psi}\sin\gamma]$, γ is the rotation angle around axis x in $\dot{\gamma}$.Thus

$$\begin{bmatrix} \omega_x \\ \omega_y \\ \omega_z \end{bmatrix} = B_A \begin{bmatrix} 0 \\ 0 \\ \dot{\varphi} \end{bmatrix} + \begin{bmatrix} 0 \\ \dot{\psi}\cos\gamma \\ -\dot{\psi}\sin\gamma \end{bmatrix} + \begin{bmatrix} \dot{\gamma} \\ 0 \\ 0 \end{bmatrix} \tag{23}$$

The corresopnding scalar format is

$$\begin{cases} \omega_x = -\dot{\varphi}\sin\psi + \dot{\gamma} \\ \omega_y = \dot{\varphi}\cos\psi\sin\gamma + \dot{\psi}\cos\gamma \\ \omega_z = \dot{\varphi}\cos\psi\cos\gamma - \dot{\psi}\sin\gamma \end{cases} \tag{24}$$

Describing Eq.(24) in another form

$$\begin{cases} \dfrac{d\varphi}{dt} = \dfrac{1}{\cos\psi}(\omega_y\sin\gamma + \omega_z\cos\gamma) \\ \dfrac{d\psi}{dt} = \omega_y\cos\gamma - \omega_z\sin\gamma \\ \dfrac{d\gamma}{dt} = \omega_x + \tan\psi(\omega_y\sin\gamma + \omega_z\cos\gamma) \end{cases} \tag{25}$$

Eq.(25) is the angular equations of motion.

3 Analysis for an example

To make analysis simple, some hypothesis is necessary to define ahead,for example:

a) The aircraft is composed of three components: fuselage、left moving wing and right moving wing.

b) All wings have no aileron.

c) Fuselage(including fixed wing part) maintains relative fixed position.

d) All wings have no dihedral.

e) Stretching rate of telescopic wing is some contant value.

f) Mass of left and right wing is $\boldsymbol{m_1}$ and $\boldsymbol{m_2}$, supposing that $\boldsymbol{m_1}=\boldsymbol{m_2}$.

Based on the above hypothesis,a relative telescopic wing dynamic model would be established.While the wings are in the retractile status,there is one point $\boldsymbol{O}$ locating on symmetry surface of fuselage,where

$$S_{ox} = \int_\Omega r'_{ix}dm = 0, S_{oy} = \int_\Omega r'_{iy}dm = 0$$

Ω represents the integral of whole aicraft.As is shown in Fig.5,stretching rates of left and right wing are $\boldsymbol{v_1}$,$\boldsymbol{v_2}$, respectively. Sweep angle is χ.At the moment, only $S_{oy} = \int_\Omega r'_{iy}dm = 0$.

Then,

$$S_{ox} = m_1(d_2 - d_1)\sin(-\chi), S_{oz} = m_1(d_2 - d_1)\cos\chi \tag{26}$$

Where $d_1 = v_1 t, d_2 = v_2 t$.

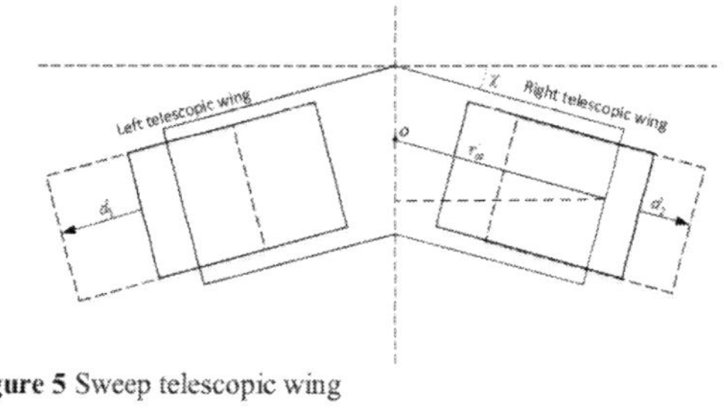

Figure 5 Sweep telescopic wing

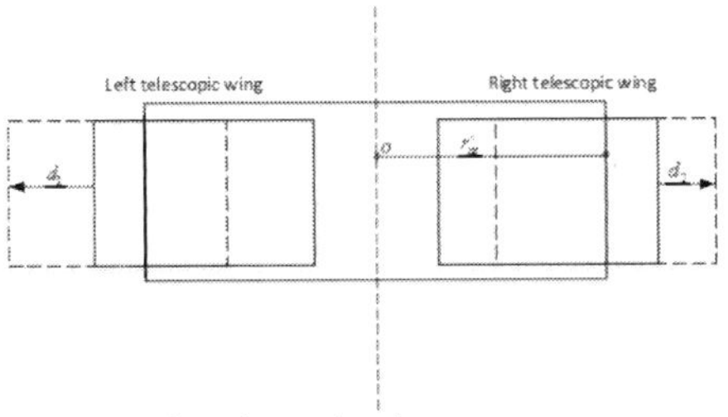

Figure 6 Rectangular telescopic wing

Futhrmore ,accoring to the feature of rectangular telescopic wing , Eq.(26) can be expressed as

$$S_{ox}=0, S_{oz}=m_1(d_2-d_1)=m_1(v_2-v_1)t \tag{27}$$

Dynamic and kinematic equations are simplified as

$$\begin{cases} m(\dot{V}_x+\omega_y\cdot V_z-\omega_z\cdot V_y)+\dot{\omega}_y\cdot S_{oz}+2\omega_y\cdot \dot{S}_{oz} \\ +\omega_x\omega_z S_{oz}=F_x \\ m(\dot{V}_y+\omega_z\cdot V_x-\omega_x\cdot V_z)+\dot{\omega}_x\cdot S_{oz}+2\omega_x\cdot \dot{S}_{oz} \\ +\omega_y\omega_z S_{oz}=F_y \\ m(\dot{V}_z+\omega_x\cdot V_y-\omega_y\cdot V_x)-\omega_x^2 S_{oz}-\omega_y^2 S_{oz} \\ =F_z \end{cases} \tag{28}$$

$$\begin{cases} I_x\cdot\dot{\omega}_x+\dot{I}_x\cdot\omega_x+(\omega_y(I_z\cdot\omega_z)-\omega_z(I_y\cdot\omega_y)) \\ +S_{oz}\dot{V}_y-\dot{S}_{oz}(V_x\cdot\omega_z-V_z\cdot\omega_x)=M_{ox} \\ I_y\cdot\dot{\omega}_y+\dot{I}_y\cdot\omega_y+(\omega_z(I_x\cdot\omega_x)-\omega_x(I_z\cdot\omega_z)) \\ +S_{oz}\dot{V}_x-S_{oz}(V_y\cdot\omega_z-V_z\cdot\omega_y)=M_{oy} \\ I_z\cdot\dot{\omega}_z+\dot{I}_z\cdot\omega_z+(\omega_x(I_y\cdot\omega_y)-\omega_y(I_x\cdot\omega_x)) \\ =M_{oz} \end{cases} \tag{29}$$

4 Estimation of aerodynamic force

The theoretical value of thin airfoil lift line slope is 2π, the value of thick wing increases with relative thickness of airfoil increasing.Experiment value is about $0.9\times 2\pi$, the correction value can approximately be expressed as

$$C_l^\alpha=1.8\pi(1+0.8c) \tag{30}$$

Where c is thickness of airfoil .Because of tip effect, the slope of wing lift line is lower than airfoil ,

$$C_L^\alpha=\frac{C_l^\alpha}{1+C_l^\alpha/(\pi\lambda e)} \tag{31}$$

Where C_L^α, C_l^α is the slope of wing and airfoil lift line, respectively, λ is aspect ratio, e is Oswald dfficiency factor, which is a function of aspect ratio and sweep angle. For rectangular wing,

$$e=1.78(1-0.045\lambda^{0.68})-0.64 \tag{32}$$

Drag coefficient of wing can be estimated that[14]

$$C_D=C_{D0}+C_{Di} \tag{33}$$

$$C_{D0}=C_f(S_W/S), C_{Di}=KC_L^2 \tag{34}$$

Where C_f is equivalent skin friction coefficient, S_W is wetted area of wing ,S is reference area of wing,K is damping factor by lift. At low velocity, equivalent skin friction coefficient of wing in turbulent flow can be approximately caculated as

$$C_f=\frac{0.455}{(\lg\mathrm{Re})^{2.58}} \tag{35}$$

Usually ,

$$K=\frac{1}{\pi\lambda e} \tag{36}$$

For some specific rectangular telescopic wing, lift and drag force can be obtained from above formulae , then it is necessary to project them on inertial axes, the following is transformation matrix.

$$B_V=\begin{bmatrix} \cos\alpha\cos\beta & \sin\alpha & -\cos\alpha\sin\beta \\ -\sin\alpha\cos\beta & \cos\alpha & \sin\alpha\sin\beta \\ \sin\beta & 0 & \cos\beta \end{bmatrix}$$

Where α (defined as angle of attack) is the angle between the axis ox and velocity' projection on $x_A oz_A$ surface, β (defined as angle of sideslip) is the angle between velocity vetor and $x_A oz_A$ surface.

5 Analysis of roll performance

5.1 Rolling moment coefficient generated by aileron is caculated

Increment of rolling moment coefficient by aileron deflection

$$\mathrm{D}m_x=\frac{\mathrm{D}M_x}{qSb}=\frac{C_l qcydy}{qSb}=\frac{C_l cydy}{Sb} \tag{37}$$

Where b is maximum wing span, S is maximum wing area, section lift coefficient of left wing can be given as

$$C_l=C_l^\alpha\frac{da}{dd_a}d_a=C_l^\alpha t_a d_a \tag{38}$$

τ_a is aileron dfficiency factor, which represents equivalent angles of attack of wing section with aileron when aileron deflects1 degree, and can be estimated by $\tau_a=(c_a/c)^{0.5}$.

Where c_a is chord of aileron.For rectangular wing, chord of wing c is constant, consdering a couple of ailerons , through integral of Δm_x,

$$m_x=\frac{C_L^\alpha t_a}{Sb}(y_2^2-y_1^2)d_a \tag{39}$$

Where C_L^αandt_a have been corrected by considering effect of practical flow, y_1 and y_2 is the starting and ending position of aileron along span .

5.2 Rolling moment coefficient by asymmetric wing can be caculated

Rolling moment coefficient:

$$m_x = \frac{C_L^a a S_a \Delta L}{2Sb} \quad (40)$$

Where ΔL is difference value of left and right wing.*Sa* is the actual wing area. Fixed wing is 2m, moving wing span is 1m, chord is 1m.If aileron is installed in the 25% of chord and its length takes up the whole moving wing span. NACA LS 0417 is selected as airfoil, then C_l^a =2.0448π, S=6m², b=6m, τ_a=0.5.The results of roll moment coefficient are shown in Fig.7 and Fig.8.

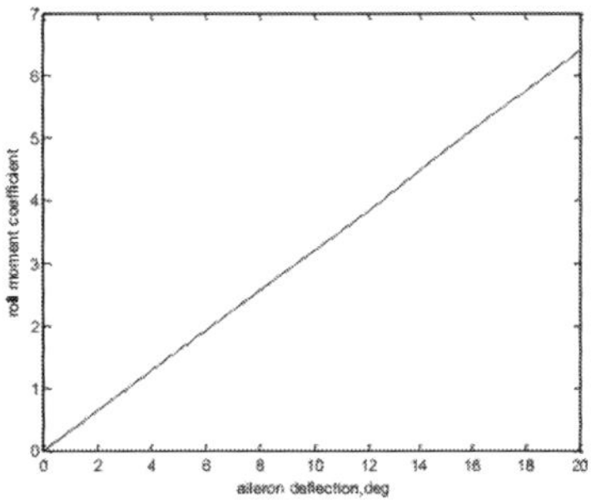

Figure 7 Rolling moment coefficient by aileron

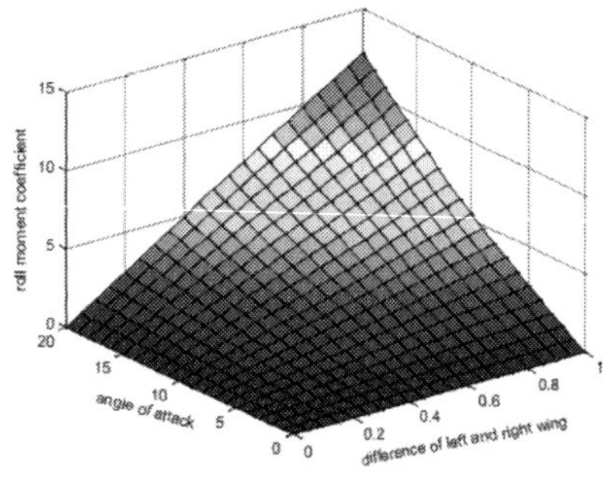

Figure 8 Rolling moment coefficient by telescopic wing

It can be seen from Fig.7 and Fig.8 that rolling moment coefficient generated by aileron is linear change , while that by elescopic wing is non-linear. When rolling moment coefficient of aircraft in a large angle of attack is easier controlled than in a small angle of attack.Meanwhile, rolling moment coefficient increases with ΔL increasing.

6 Conclusions

In this paper, the dynamic equations and kinematic equations of asymmetric telescopic wing aircraft are developed.From the equations, it is very complex and difficult to solve such equation series , because several parameters change with time.Hower, the model is versatile for any telescopic morphing aircraft, in some simple assumptions. This paper also bulid the model of rectangular telescopic wing with no sweep and dihedral.Some estimation formulae are applied to caculate the aerodynamic force . Results show that at low velocity high aspect ratio wing has better aerodynamic performance than low aspect ratio,while at high velocity low aspect ratio wing has superior performance. So, telescopic wing can enhance overall fight performance for a large velocity range.

The roll perfoamance also is evaluated by asymmetric wing span, aircraft can generate roll moment by change the stretching length of left and right wing. Results shows that roll moment coefficient varys evidently with difference value of telescopic wings' length and angle of attack. It is easy to controll the roll rate in this way. Moreover, roll performance of asymmetric wing is non-linear change. In the case of large angle of attack and ΔL, telescopic wings behave better roll performance.

References

1. A., W. T. "Morphing Aircraft Systems : Historical Perspectives and Future Challenges," *Journal of Aircraft* Vol. 50, No. 2, pp. 337-350(2013)
2. Béguin, B., Breitsamter, C., Adams, N., et al. "Aerodynamic Investigations of a Morphing Membrane Wing," *AIAA Journal* Vol. 50, No. 11, pp. 2588-2599(2012)
3. Tomohiro, Y., Aya, S., and Yoshiyasu, H. "Development and Wind Tunnel Test of Variable Camber Morphing Wing," *22nd AIAA/ASME/AHS Adaptive Structures Conference*. American Institute of Aeronautics and Astronautics, National Harbor, Maryland, (2014).
4. Vale, J., Leite, A., Lau, F., et al. "Aero-Structural Optimization and Performance Evaluation of a Morphing Wing with Variable Span and Camber," *Journal of Intelligent Material Systems and Structures* Vol. 22, No. 10, July 1, pp. 1057-1073,(2011)
5. Pecora, R., Amoroso, F., and Lecce, L. "Effectiveness of Wing Twist Morphing in Roll Control," *Journal of Aircraft* Vol. 49, No. 6, November-December, pp. 1666-1674,(2012)
6. Spearman, M. L., and Langley, N. "Aerodynamic Research at NACA / NASA Langley Related to the Use of Variable-Sweep Wings," *50th AIAA Aerospace Sciences Meeting including the New Horizons Forum and Aerospace Exposition*. Nashville, Tennessee, (2012)
7. Hall, J. "Investigation of Variable Wing-Sweep for Applications in Micro Air Vehicles," *Infotech@Aerospace*. Arlington, Virginia, (2005)
8. Bae, J.-S., Seigler, T. M., and Inman, D. J. "Aerodynamic and Static Aeroelastic Characteristics of a Variable-Span Morphing Wing," *Journal of Aircraft* Vol. 42, No. 2, March-April, pp. 528-534,(2005)
9. Junfu, L., Junqiang, A., Shitu, L., et al. "Research on Telescopic Wing Morphing Technology," *Advances in Aeronautical Science and Engineering* Vol. 3, No. 1, pp. 2-5,2012 [10] Dongping, J., and Wenmei, Z. "Longitudinal flight dynamics of morphing aircraft with telescopic wings," *Chinese Journal of Applied*

Mechanics Vol. Vol.31 No.3, pp. 468-472,(2014)
11. Adam, N., John, V., Leland, C., et al. "Morphing Aircaft Dynamical Model: Longitudinal Shape Changes," *AIAA Atmospheric Flight Mechanics Conference and Exhibit*. American Institute of Aeronautics and Astronautics, Honolulu, Hawaii, (2008)
12. Peiran, J., Kejun, C., and Li, H. *Ballistics of Remote Rocket(In Chinese)*. China Changsha National University of Defense Technology Press, (1993)
13. Zhenping, F., Wanchun, C., and Shuguang, Z. *Flight Dynamics of Aerial Vehicle(In Chinese)*. China Beijing: Beihang University Press, (2005)
14. Adam, N., John, V., Thomas, S., et al. "Six Degree-of-Freedom Dynamical Model of a Morphing Aircraft," *AIAA Atmospheric Flight Mechanics Conference*. American Institute of Aeronautics and Astronautics, Chicago, Illinois, (2009)

Improved Mask Protected DES using RSA Algorithm

S.Asha Latha[1], A.Sivabalan[2]
[1]*Research Scholar, Sathyabama University,Chennai- 600119*
[2]*Manager Research, NEC Mobile networks Excellence Centre, Chennai.*
asha_26_2001@yahoo.com, sivabalan.armugam@necindia.in

Abstract: The data encryption standard is a pioneering and farsighted standard which helped to set a new paradigm for encryption standards. But now DES is considered to be insecure for some application. Asymmetric mask protected DES is an advanced encryption method for effectively protecting the advanced DES. There are still probabilities to improve its security. This paper propose a method, which introduce a RSA key generation scheme in mask protected DES instead of plain key, which result in enhancement in the security of present asymmetric mask protected DES. We further propose a Vedic mathematical method of RSA implementation which reduce the complexity of computation in RSA block thereby resulting in reduced delay (four times)that improves the performance of overall system. The software implementation was performed using Xilinx 13.2 and Model-Sim was used for the simulation environment.

Keywords: cryptography, security, data confidentiality, RSA, DES, Vedic mathematics

I Introduction

Cryptography is usually referred to, as the study of securing information.The aim of cryptography is not to hide the existence of a message but it rather hides its meaning. Encryption is the process of converting plain text to cipher text. The process of converting cipher text back to plain text is called decryption[4],[7].There are two basic cryptographic techniques one in symmetric or private key cryptography and another is asymmetric key cryptography . Data Encryption Standard (DES) is a widely-used method of data encryption using a private (secret) key that was judged so difficult to break[9]. Most symmetric encryption schemes today are based on this structure. Although this is considered "strong" encryption, DES is one of the symmetric encryption algorithm, which is considered to be insecure now a days. Asymmetric mask protected DES is also a symmetric cryptography where an extra masking technique is used to increase the security of DES shown in fig 1 and fig 2. There are still probabilities to improve its security and efficiency. Here, a new method is introduced where in the already existing key generation scheme is replaced by RSA encryption. RSA prime factorization involved in it increases the security thus making it resistant to side channel attack such as correlation power analysis attacks and differential power analysis attack[2],[3]which is used to crack the plain text[10],[11]. Further Vedic mathematical calculations are involved to reduce the complexity of encryption in RSA block which further improves the efficiency of the proposed method[13].The rest of the paper is organized as follows. Section II describes the asymmetric mask protected DES, and Vedic Implementation. Section III describes the proposed improved encryption mechanism using RSA algorithm. Section IV describes the simulation environment and results. Section V contains conclusion and the future scope.

II. BACKGROUND

a. Asymmetric Mask Protected DES

The asymmetric mask protected DES is a data encryption standard where a masking technique is introduced in the normal DES algorithm[1]. Like most of the encryption schemes mask protected DES expects two inputs – the plain text to be encrypted and the secret key. The main idea in this method is to add different random numbers in the first, last and internal rounds. This method also has 16 rounds of operation similar to DES. Hence multiple mask operations added at different moments and locations helps to break the relation between power consumption and key thereby improving security.

b. Vedic Implementation

The RSA Algorithm is the most popular asymmetric key cryptographic algorithm.. The RSA Algorithm is based on the mathematical functions that are easy to find and multiply the large numbers together, but it is extremely difficult to factor their product[5]. The public and private keys in RSA are based on very large numbers. To increase the computation speed with minimized hardware the Vedic mathematics [6]multiplication principle is used. These architectures are used to improve the speed of the RSA algorithm with reduced hardware[7],[8].

Mask Protected DES Encryption and Decryption:

ENCRYPTION

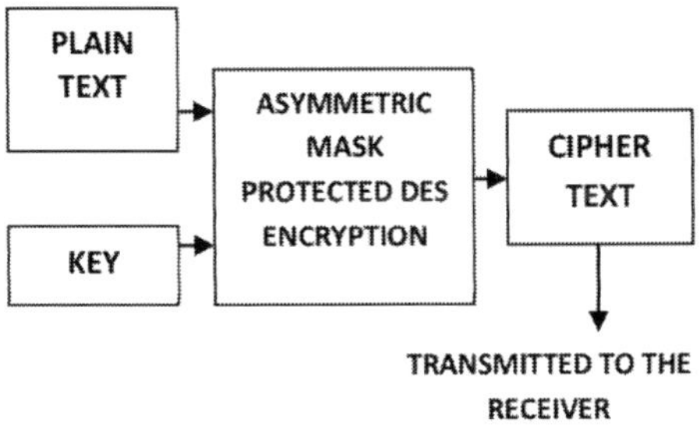

Fig 1. Mask protected DES Encryption

DECRYPTION

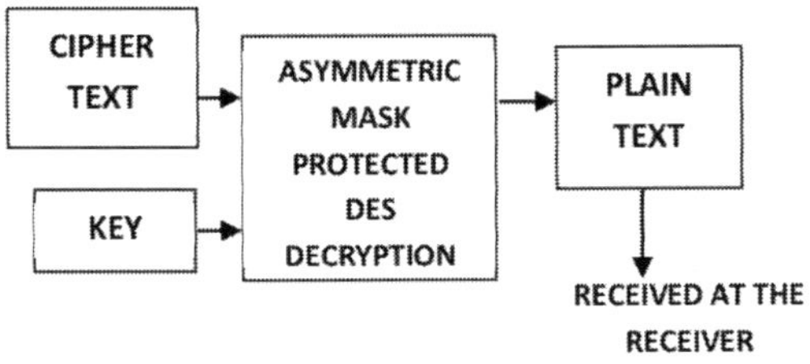

Fig 2. Mask Protected DES Decryption

III Improved Asymmetric Mask Protected DES

In the proposed system, instead of normal key generation of mask protected DES, RSA algorithm which is a asymmetric key algorithm is used. The advantage of asymmetric key is that it is secure from the middle attack unlike symmetric key. During the process of encryption, asymmetric mask protected DES encrypts the plain text to produce the cipher text.The key is encrypted using RSA at the sender side before transmitting it to the receiver shown in fig 3. Finally the combination of the key from RSA encryption and the cipher text from mask protected DES is sent out. At the receiver two cipher texts are obtained shown in fig 4. One is cipher text from asymmetric mask protected DES encryption and the other is the cipher key from the RSA encryption. The receiver decrypts the cipher key received from RSA encryption by their own private key. Then using the decrypted original key, the cipher text is decrypted. Finally the original plain text is received at the receiver end. This method ensures improved data confidentiality and integrity because of the deal protection of asymmetric mask protected DES algorithm and RSA algorithm, the data in transit is safe.

Improved Mask Protected RSA Encryption:

ENCRYPTION

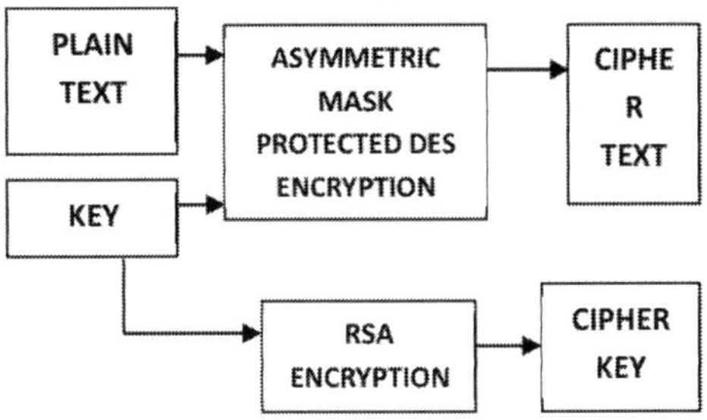

Fig 3.Improved Mask Protected RSA Encryption

Improved Mask Protected RSA Decryption:

DECRYPTION

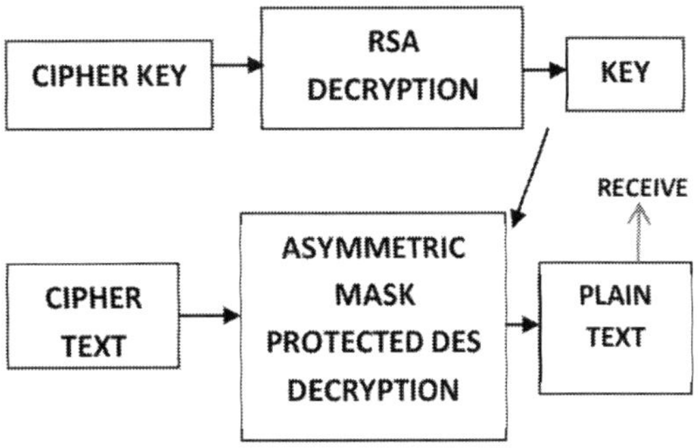

Fig 4.Improved Mask Protected RSA

DOI: 10.1051/matecconf/20166501005

On other hand, one has to accept the fact that securing always increases the complexity of an encryption procedure thereby consuming more power for encryption and decryption. Secondly this complexity reduces the speed of data processing thereby increasing the delay. Hence security as always compromised with speed and throughput performance to some extent in any encryption techniques. In the proposed method the existence of array multiplier is replaced by Vedic multiplier in the RSA block. This replacement produced four times reduced delay values when compared to the usual array multiplier used in RSA. To increase the computation speed, the multiplication principle of Vedic mathematics is used. And also "URDHVA-TIRYAKBHYAM" is the sutra (principle) which is used to compute the multiplication. The significance of this technique is that it computes the partial products in one step and avoids shifting operation which reduces the complexity of the algorithm and saves both the execution time and hardware. This ultimately improves the speed of RSA algorithm[13],[12]. Hence the secure dual protection of asymmetric mask protected DES and RSA is achieved with less complexity.

IV. Result and discussion

The proposed method is simulated using ModelSim 6.3 g and code in written in verilog using Xilinx. Fig.5 shows the encrypted output of proposed method. The input and output details are given below.

Plaintext -
0000000000010001001000100011001101000100010101010110011001110111

InputKey-
0000000100100010001101000101011000000001001000100011010001010110

Mask-0000000100100010001101000101011

Ciphertext-
1000100011100001101000101001100110010010001100100111010011000010

Cipherkey-
000000000000000000000000000000010000000000000000000000100110101000000000000000000000000001000
0110100000000000000000000000001100101010000000
000000000000000000000001000000000000000000000
0010011010100000000000000000000000000100001101000
0000000000000000000000110010101

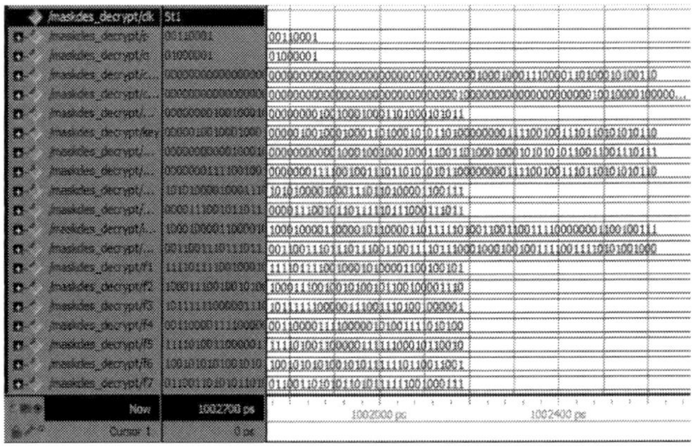

Fig.5 Encrypted output waveform for proposed system

Fig.6 shows the decrypted output of proposed method. The input and output details are given below.

Ciphertext-
1000100011100001101000101001100110010010001100100111010011000010

Cipherkey-
000000000000000000000000000000010000000000000000000000100110101000
00000000000000000000100001101000000000000000000
000000001100101010000000000000000000000000000000
00010000000000000000000000001001101010000000000
000000000000010000110100000000000000000000000000
0110010101

Mask-0000000100100010001101000101011

Plaintext-
0000000000010001001000100011001101000100010101010110011001110111

InputKey-
0000000100100010001101000101011000000001001000100011010001010110

Even if the attack sample is increased five times, the attacker still cannot gain the key when improved asymmetric masked DES using RSA is used. Hence the proposed method is resistant to differential power analysis attack and correlation power analysis attack.

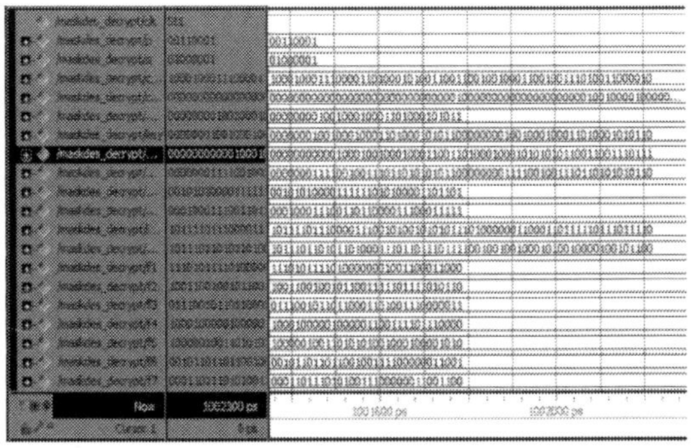

Fig.6 Decrypted waveform for proposed system

V. CONCLUSION

The improvedmask protected DES encryption mechanism using RSA algorithm is a full featured circuit including key generation, data encryption and data decryption. Each sub component and top module of the proposed encryption mechanism was implemented using verilog and was simulated using ModelSim which proved to be functionally correct. Dual security is achieved by the use of random numbers and RSA algorithm thus resistant to side channel attack. The complexity in the proposed system is reduced by using Vedic mathematical calculation which produced four times lesser delay values. Since the demand for higher levels of security is increasing the use of the proposed mechanism makes the entire algorithm difficult to crack. As a result, security is enhanced. The obtained Less delay values promotes high speed and low power consumption, thus promoting efficient hardware implementation on FPGAs. Future work can be carried out on cadence. The proposed method can also be implemented in Bluetooth technology and hybrid encryption. The work can also be extended to large bits such as 256 or 1024 or even longer and can be used in many applications where security is of major concern.

References:

1. Weiwei Shan Member IEEE "Evaluation of Correlation Power Analysis Resistance and Its Application on Asymmetric Mask Protected Data Encryption Standard Hardware", IEEE Transaction on instrumentation and measurement, Vol. 62 No.10 (2013).
2. Akkar.M.L and Giraud .C (2001), "An implementation of DES and AES secure against some attacks", in Proc. Cryptographic Hardware Embedded Syst, Paris, France, pp.309-318.
3. Weiwei Shan, Xin Chen, Bo Li, Peng Cao, Jie Li, Gugang Gao, and Longxing Shi (2013), "Evaluation of Correlation Power Analysis Resistance and Its Application on Asymmetric Mask Protected Data Encryption Standard Hardware", IEEE Transactions on Instrumentation and Measurement, Vol. 62, No.10, pp.2716-2724.
4. WilliamStallings (2005), "Cryptography and Network Security Principles and Practices", Prentice Hall, November 16, pp.24-28, 56-83.
5. HimanshuThapliyal and M.B Srinivas, "VLSI Implementation of RSA Encryption System Using Ancient Indian Vedic Mathematics", Center for VLSI and Embedded System Technologies, International Institute of Information Technology Hyderabad-500019, India.
6. Shamim Akhter, "VHDL Implementation of Fast NXN Multiplier Based on Vedic Mathematics", Jaypee Institute of Information Technology, University, Noida, 201307 UP, INDIA, 2007 IEEE.
7. BehrouzAForouzan, DebdeepMukhopadhyay,"Cryptography and Network Security", second edition, *Tata McGraw Hill Education Pvt Ltd*, 2008.
8. Nitish Aggarwal, KartikAsooja, SaurabhShekhar Verma, SapnaNegi," An Improvement in the Restoring Division Algorithm", IEEE 2009.
9. Sumit Vaidya, Deepak Dandekar, "Delay-Power Performance Comparison of multipliers in VLSI circuit design", *International Journal of Computer Networks & Communications* (IJCNC), Vol.2, No.4, July 2010.
10. J. Wu, Y. Shi, and M. Choi, "Measurement and evaluation of power analysis attacks on asynchronous S-box," *IEEE Trans. Instrum.Meas.* vol. 61, no. 10, pp.2765–2775, 2012.
11. J. Li, Y. Lv, H. Sun, and W. Shan, "A power analysis resistant DES cryptographic algorithm and its hardware design," in *Proc. 3rd Int. Conf. Digit. Manuf. Autom.*, GuiLin, China, pp. 121–124, 2012.
12. Chiranth E, Chakravarthy H.V.A, Nagamohanareddy P, Umesh T.H, Chethan Kumar M, "Implementation of RSA Cryptosystem Using Verilog", International Journal of Scientific & Engineering Research Volume 2, Issue 5, May-2011
13. Bhaskar, Ganapathi Hegde, P.R.Vaya, "Anefficient hardware model for RSA Encryption

system usingVedic mathematics", International Conference on Communication Technology and SystemDesign2011.

Imperfect Preventive Maintenance Model Study Based On Reliability Limitation

Qian Zhou[1, a], Yufeng Sun[2]

[1]*Technology Center, AVIC CHENGDU AIRCRAFT INDUSTRIAL(GROUP) CO.,LTD , Chengdu 610073, China*
[2]*School of Reliability and Systems Engineering, Beihang University, Beijing 100191, China*

Abstract. Effective maintenance is crucial for equipment performance in industry. Imperfect maintenance conform to actual failure process. Taking the dynamic preventive maintenance cost into account, the preventive maintenance model was constructed by using age reduction factor. The model regards the minimization of repair cost rate as final target. It use allowed smallest reliability as the replacement condition. Equipment life was assumed to follow two parameters Weibull distribution since it was one of the most commonly adopted distributions to fit cumulative failure problems. Eventually the example verifies the rationality and benefits of the model.

Introduction

Equipment is the major asset of the modern enterprise. Preventive maintenance is an ideal repair method in eliminating fault. The traditional repair assume that the equipment will return to perfect state or minimal repair state after repair, which is not consistent with the actual situation. Most of the equipment will restore to a state between them, which is called imperfect maintenance[1]. For example, the proper adjustment of the engine is imperfect maintenance, because the engine will not return to a new state but a state improved a lot after the adjustment.

Pham and Wang divided imperfect maintenance into eight categories: (p,q) rule, $(p(t),q(t))$ rule, improvement factor method, the virtual age method, impact model method, (α,β) rule, composite(p,q)rules and other[2]. Malik proposed the concept of improvement factor[3]. It can measure age improvement degree and failure rate after repair. Chan and Shaw gives two types of failure rate recovery mode that depends on the length of service and maintenance effect[4]: fixed reduced type and ratio of reduced type.

The traditional preventive maintenance method calculates the maintenance cycle according to the average life, reliability or single repair cost. It doesn't consider the balance between reliability and repair cost. By introducing the reliability constraints, the paper establishes a better preventive maintenance model considering both cost and reliability.

Imperfect preventive maintenance model

The following are several basic assumptions of the model:

(1) Equipment conduct (N-1)th imperfect maintenance then it will be replaced in the Nth in the infinite time domain.

(2) Equipment will conduct minimum repair in the periodic preventive maintenance interval h if failure occur. The minimum repair can only make equipment return to a normal operation state, it doesn't change failure rate.

(3) The effect of imperfect preventive maintenance can be achieved by age reduction factor. Age reduction factor is influenced by equipment age and preventive maintenance cost rate. It is a dynamic value.

(4) Single preventive maintenance cost shows a linear relationship with repair numbers. It is expressed as $c_{pmi} = c_f + ic_v$ .Equipment is worn with increase of device age, each maintenance time becomes longer, which is $t_{pmi} = (1/k)\cdot i\cdot h$, $1/k$ is time adjustment parameter.

(5) Single minimal repair cost c_{mr} and single repair time t_{mr} are constant. Production loss cost per unit time c_s is constant.

Figure.1 is the periodic imperfect preventive maintenance strategy model diagram.

Figure 1. imperfect preventive maintenance model

Equipment operation is a degradation process. Imperfect preventive maintenance can improve the life of

Qian Zhou: zhouqian826@126.com

equipment. Age reduction factor η_i is used to indicate the recovery degree after the preventive maintenance [5].

$$\eta_i = \left(a \cdot \frac{c_{pmi}}{c_{pr}} \right)^{b \cdot i}, i = 1, \ldots, N-1 \qquad (1)$$

a is repair cost adjustment factor, and $0 \le a \le c_{pr}/c_{pmi}$. c_{pmi} is single preventive maintenance cost. c_{pr} is one replacement cost, and $c_{pmi} \le c_{pr}$. b is adjustment coefficient of repair times, and $0 < b < 1$.

We can use age reduction factor to obtain the effective age. The bigger the age reduction factor is, the smaller effective age is. This paper uses the age proportion reduced model, only the maintenance of the last age can get the percentage recovery. The effective age before and after the maintenance is expressed as formula (2).

$$T_1^- = h, T_1^+ = (1 - \eta_1) h \qquad (2)$$

Similarly, the effective age before and after the ith maintenance is shown in formula (3).

$$T_i^- = (i - \sum_{j=1}^{i-1} \eta_j) h, T_i^+ = (i - \sum_{j=1}^{i} \eta_j) h \qquad (3)$$

After implementation of $(N-1)^{th}$ prevention maintenance, replacement will be implemented in the Nth maintenance cycle. The preventive replacement timeline is shown in formula (4).

$$T_N^- = (N - \sum_{j=1}^{N-1} \eta_j) h \qquad (4)$$

The Weibull distribution form can describe the failure regularity of mechanical equipment, it is widely used in practical production. The failure rate expression of the distribution is shown in formula (5). β is the shape parameter. α is a scale parameter.

$$\lambda(t) = \frac{\beta}{\alpha} \left(\frac{t}{\alpha} \right)^{\beta - 1} \qquad (5)$$

Equipment failure numbers w_i in the ith imperfect maintenance cycle can be obtained according to the failure rate.

$$\begin{aligned} w_i &= \int_{T_{i-1}^+}^{T_i^-} \lambda(t) dt \\ &= \left(\frac{h}{\alpha} \right)^{\beta} \left[\left(i - \sum_{j=1}^{i-1} \eta_i \right)^{\beta} - \left(i - 1 - \sum_{j=1}^{i-1} \eta_i \right)^{\beta} \right] \end{aligned} \qquad (6)$$

Total costs C_{total} include total preventive repair costs C_{pm}, total minimum repair costs C_{mr}, production loss costs C_p and replacement costs C_{pr}.It is expressed as formula (7).

$$\begin{aligned} C_{total} &= C_{pm} + C_{mr} + C_p + C_{pr} \\ &= \sum_{i=1}^{N-1} c_{pmi} + \sum_{i=1}^{N} c_{mr} w_i + c_s \left(\sum_{i=1}^{N-1} t_{pmi} + \sum_{i=1}^{N} t_{mr} w_i \right) + C_{pr} \\ &= \sum_{i=1}^{N-1} (c_f + i c_v) + \sum_{i=1}^{N} c_{mr} w_i + c_s \left(\sum_{i=1}^{N-1} \frac{i}{k} \cdot h + \sum_{i=1}^{N} t_{mr} w_i \right) + C_{pr} \\ &= (c_{mr} + c_s t_{mr}) \sum_{i=1}^{N} w_i + \left[(N-1) c_f + \frac{N(N-1) c_v}{2} \right] + c_s \frac{N(N-1) h}{2k} + C_{pr}. \end{aligned} \qquad (7)$$

Equipment total running time T before replacement includes preventive maintenance interval h and preventive maintenance time. The repair cost rate $C(h,N)$ is expressed as formula (8). It is a Concave function. This paper omits the proof process.

$$\begin{aligned} C(h,N) &= \frac{C_{total}}{T} \\ &= \frac{(c_{mr} + c_f t_{mr}) \sum_{i=1}^{N} \left(\frac{h}{\alpha} \right)^{\beta} \left[\left(i + 1 - \sum_{j=1}^{i} \eta_{j-1} \right)^{\beta} - \left(i - \sum_{j=1}^{i} \eta_{j-1} \right)^{\beta} \right] + \left[(N-1) c_f + \frac{N(N-1) c_v}{2} \right] + c_s \frac{N(N-1) h}{2k} + C_{pr}}{Nh + \frac{N(N-1) h}{2k}}. \end{aligned} \qquad (8)$$

This paper adopts periodic imperfect preventive strategies under failure rate limits. It is time to conduct preventive replacement when the equipment reliability is equal to the predetermined limit level $R_{\min}$. Equipment failure process obeys Weibull distribution. T_N^- is replacement time. Then h can be obtained through the following methods.

$$R(T_N^-) = 1 - F(T_N^-) = \exp\left[-(T_N^- / \alpha)^{\beta} \right] = R_{\min} \qquad (9)$$

$$h = \frac{\alpha (-\ln R_{\min})^{1/\beta}}{N - \sum_{j=1}^{N-1} \eta_j} = \frac{\alpha (-\ln R_{\min})^{1/\beta}}{N - \sum_{j=1}^{N-1} \left(a \cdot \frac{c_{pmi}}{c_{pr}} \right)^{b \cdot i}}. \qquad (10)$$

Set $N = 1$, get corresponding h , then get $C(h,N)$; set $N = N+1$; If $C(h,N) > C(h,N+1)$, record $C(h,N+1)$ as the minimum new cost.

Verification Example

Equipment fault time obey the Weibull distribution. The parameter is shown in Table 1.

Table 1. model parameter

parameter	value	parameter	value
α	221	c_f	5000
β	3	c_v	100
b	0.002	k	500
c_{pr}	4000000	$R_{\min}$	0.6
c_s	5000	c_{mr}	8000
t_{mr}	0.4		

Age reduction factor curve is shown in Figure.2. Age reduction factor decreases with increasing device age gradually. The age reduction factor of first imperfect preventive maintenance is η_1=0.9868 .while the 50th age reduction factor is η_{50}=0.5493 , the equipment only can restore to the state of 54.93%.

Using iterative algorithm I obtain the maintenance cost of different repair numbers. The relationship between repair cost rate and the maintenance numbers is shown in Figure.3. Repair cost rate decrease at first with the increase of N but it starts to increase in the end. This is consistent with the actual situation. It gets the minimum value when $N = 13, h = 89.8622$ .The maintenance interval h which changes with repair numbers is shown in Figure.4.

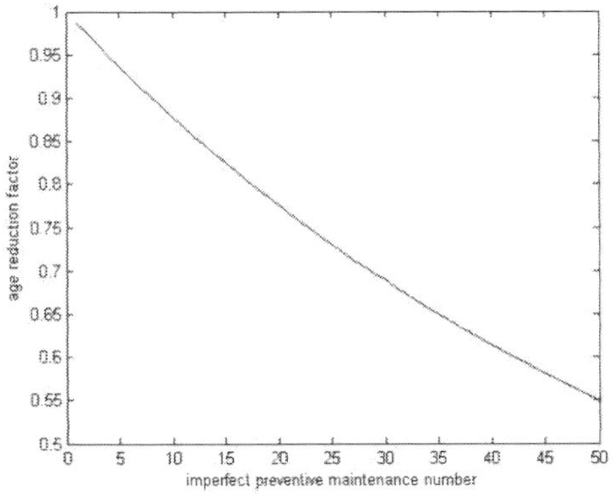

Figure 2. age reduction factor changes with N

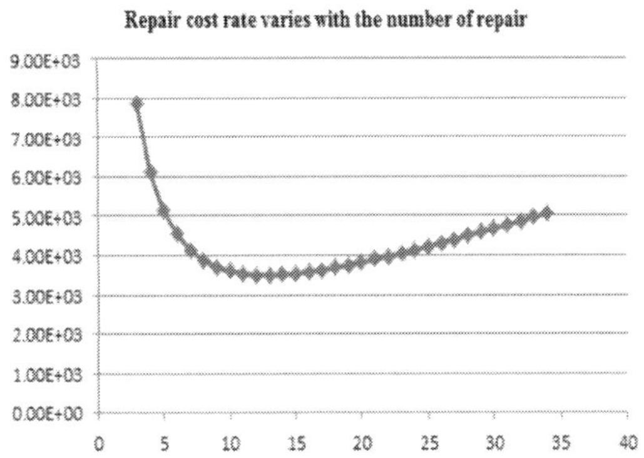

Figure 3. repair cost rate changes with N

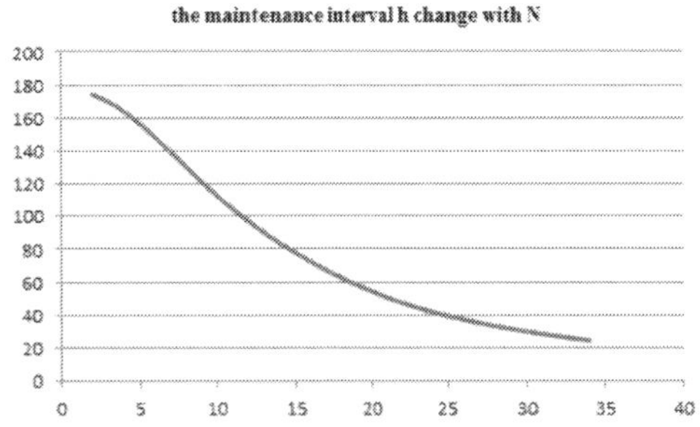

Figure 4. h changes with N

Conclusion

Equipment can't become as good as new after maintenance. The article construct imperfect preventive maintenance model based on periodic prevention. Age reduction factor is used to describe the effective age after maintenance. By adding reliability constraint condition, such policies can ensure that reliability of the equipment is kept above a certain predetermined level and thus the risk of sudden functional failure is low. The method mentioned in the article has certain reference significance on the costs optimization of enterprise. The imperfect preventive maintenance model can quantitively determine the maintenance interval and provide suggestion for the operation of the company maintenance policy.

References

1. Nakagawa T, "Optimum policies when preventive maintenance is imperfect," IEEE Transactions on Reliability, vol. R-28, Oct. 1979, pp. 331-332.
2. Pham H and Wang H. Imperfect maintenance [J].European Journal of Operational Research 94:425-438. 1996.
3. Malik, V and Jardine, A.K.S Optimal Replacement for A General Model with Imperfect Repair [J], Journal of the Operational Research Society,43/2:111-120.1992.
4. Chan, J.K. and Shaw, L. Modeling Repairable Systems with Failure Rates that Depend on Age and Maintenance [J].IEEE Transactions on Reliability, Vol.42, No.4:566-571, 1993.
5. Shi Xutian. Aging equipment preventive repair policy optimization [D].Jilin: Jilin university,2008.

MATEC Web of Conferences **65**, 01007 (2016) DOI: 10.1051/matecconf/20166501007
ICNSCM 2016

The research and application of anchor foundation

Yongping Li[1,a], Shiming Yan[2]

[1]*Northeast Dianli University, Jilin 132012, China*
[2]*Northeast Dianli University, Jilin 132012, China*

Abstract. The previous project in mountain area is often used in the ordinary rock digging foundation, the foundation of the general requirements of artificial digging, when the rock very good integrity and weathering degree is low, the time-consuming, low construction efficiency, low degree of mechanization. In order to change this situation, The study of rock bolt foundation specially set up, and the application in the engineering and the follow-up project, make it become the basic types of practical value in engineering.

1 Introduction

Tall mountain mountain and brought great difficulties to the tower foundation material transportation and construction work, the problem at the same time, but also the environmental protection. The previous project in mountain area is often used in the ordinary rock digging foundation, the foundation of the general requirements of artificial digging, when the rock very good integrity and weathering degree is low, the time-consuming, low construction efficiency, low degree of mechanization. In order to change this situation, the study of rock bolt foundation specially set up, and the application in the engineering and the follow-up project, make it become the basic types of practical value in engineering. By means of test and engineering application, we think of rock bolt foundation has the following advantages: make full use of the rock itself strength, strong ability to resist uplift; small amount of earthwork excavation, save steel and concrete;rock hole mechanized transport, less quantity, short construction period; does not destroy the original geological condition, slag less, is conducive to the Protect environment; save manpower and material resources, and has good social benefit and economic benefit, so the rock bolt foundation has very high application value in the mountains.

2 Organization of the Text

2.1 The research and analysis of Anchor foundation.

Commonly used in the transmission lines of rock anchor pile foundation type include straight anchor, bearing table, built-in type and anchor foundation, including direct anchor and bearing table were called rock bolt foundation in the engineering practice.

[a] Corresponding author: author@e-mail.org

When the rock bare rock hard breeze, can be directly called tower leg anchor bolt anchor into the rock anchor see Fig. 1. The anchor bar moderately weathered rock, using anchor into the rock, and on the anchor bar pouring pile caps, anchor bolt embedded in taichung, as shown in Fig. 2:

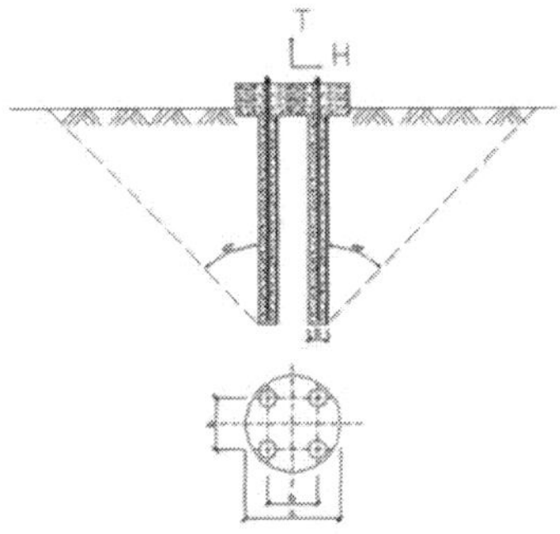

Figure 1. Vertical anchor

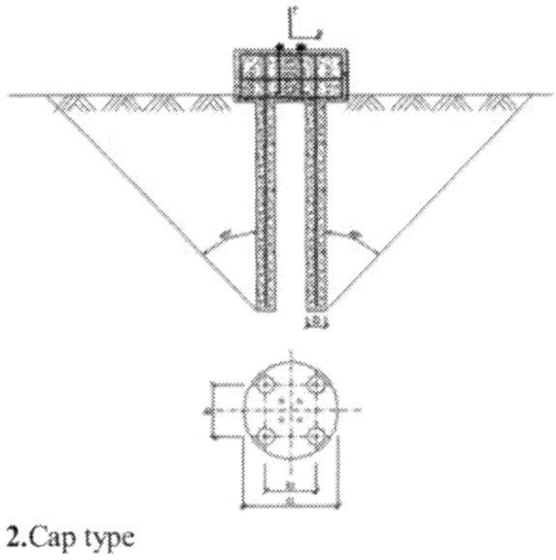

Figure 2. Cap type

Due to the high compressive bearing capacity of rock mass, so the rock bolt is generally not down load calculation, when the design, only need to meet the requirements of drawing and stable. To anchor based on assumptions when pulling damage rock mass is by the dotted line as shown in the diagram below inverted cone as a hypothetical failure surface, so in when drawing design assumptions with uniform distribution in the cone surface generation such as limit shear force on the vertical component of the sum to resist pulling force. At the same time, should guarantee anchor bolt itself of the anchor bar or the tensile strength , anchor bar and anchor bolt and mortar bond strength or bolt and rock bond strength are both meet the requirements.

In order to ensure the reliability and rationality of the anchor based on the engineering practice, organization and implementation of the rock anchor foundation experiment, and practical applications in engineering, at the same time as the rock anchor foundation to further promote the use of accumulated experience in the transmission lines.

The experiment selected the two separately according to the rock properties of different kinds of typical rock test. The rocks of two types are strongly weathered granite and moderately weathered limestone. In each geological respectively for three straight single anchor bolt pulling and two straight type of anchor bolt is basic test.

Strongly weathered granite on three different depth on the basis of the single bolt type only do pull test, the test content is: anchor bar on the stress and deformation under the action of pulling force, anchor displacement and deformation under stress, the main column, anchor displacement of surrounding rock deformation. Strongly weathered granite on two different buried depth of the straight type of anchor bolt is based on simultaneous extraction and horizontal load test, the test content is: based on the level and bolt pulling force under the action of stress and strain of the foundation anchor bar, foundation displacement and deformation of the pile caps under stress, foundation displacement of surrounding rock deformation.

In moderately weathered rock is also based on the above test items. After each test to limit load are loaded until destruction.

In strongly weathered granite trials, five of the rock anchor foundation is shear failure of rock mass, it is an ideal result for rock bolt foundation, anchor bar and bond strength of the main column and the bond strength between concrete main column and the rock mass has reached the design requirements, at this time can make full use of the shear strength of rock mass itself to resist pulling force. Test data show that when the shear failure of rock mass, tau s value has reached 47 kN/m2 ~ 58 kN/m2, this shows that: under the geological conditions of strong weathering granite gneiss using rock anchor foundation is reasonable and reliable.

In moderately weathered limestone experiment, three single anchor experiment foundation for bond damage, the main column shows the bedrock itself inherent strength is larger, and the other two socket bench test, no signs of damage in medium weathered limestone geological conditions, the bearing capacity of the rock anchor foundation is not controlled by shear failure of rock mass.

According to the test of the ultimate bearing capacity of the press "obvious turning point method" and "the ultimate load method" calculate the ultimate pull-out force, the results of numerical see the pull force test limit table. Compared with the test load data in the table, the anchor foundation has the very high reliability

under horizontal load, deformation of Rock anchor foundation is within the scope of the permit, explain rock anchor foundation has enough ability to resist horizontal load.

2.2 Economical comparison

In mountainous area, the application of the rock foundation and application of ordinary bolt foundation, anchor based not only on the construction craft is advanced, but also has more obvious advantages in the economic and social benefits. By ordinary rock excavation foundation and anchor foundation single-base all the data in the budget sheet, based in the indicators for the anchor bolt can verify all has the good economic efficiency.

Table 1. Ordinary rock excavation foundation and anchor foundation single-base budget.

cost / foundation	cost of materials		Installation fee		freight	total
	steel	concrete	construction	excavation		
digging foundation	4510	3952	11064	7295.00	7800	34621
anchor foundation	4400	505	1556	14040.00	900	21401

Note: use more straight line tower foundation in this project are compared the economic benefits.

2.3 The advantage of the anchor foundation

The design on the basis of anchor bolt embedded in rock mass directly, make full use of the wholeness and completeness of the rock mass, foundation less dosage of concrete and steel consumption, make the foundation ontology cost reductions and has high economic benefit.

Based the surface part is small, the construction base level required is small, so quantity of earth-rock prescribing rarely, also need not open rock blasting, reduces the cost of earthwork quantity, reduce the damage to the original landform mountains, is advantageous to the ecological environment protection.

Anchor drilling machine is used for the anchor foundation, realized the mechanization of mountain foundation construction, a new construction technology, construction of the characteristics of fast speed, which can effectively shorten the construction period, improve the construction quality, reduce manpower requirements at the same time, reduce the labor intensity of workers, labor protection the job.

Anchor foundation is especially suitable for tall mountains, due to the advanced construction technology, construction base is small, the material province etc, and, ordinary difficulty for the rock foundation construction, construction site materials difficult to transport.

2.4 Design of safeguard measures

Through the research, bolt foundation in highly weathered and weathered rock tower have good applicability, mountain foundation design can play to their advantage.

Because of the huge influence of weathering degree of rock bedding joints and rock on bearing capacity of foundation, the existence of large discreteness, design personnel should be based on the surface of rock cover thickness, the stability of rock mass, rock weathering degree of firmness, bedding and fractures, etc., using the combination of full range leg length of iron tower foundation design.

Designers should be technology for construction personnel work before construction, to understand the design intention and the main design principle,master the design of the technical requirements for construction.

Construction should carry out the construction in strict accordance with the design drawings. Designers should base in place by examining trough.

3 Conclusions

The use of rock bolt foundation can save concrete material quantity and the amount of excavation, reduce the engineering investment. At the same time as little as possible to destroy the original topography, Protect environment and vegetation, reflects the good social benefit and economic benefit, and compared with the ordinary rock digging foundation has the obvious superiority .

1. Ministry of Construction of the People's Republic of China, GB50011-2010 code for seismic design of buildings[S], Beijing, 2010
2. E. Savory,G.A.R. Parke,P. Disney,N. Toy.Wind-induced transmission tower foundation loads: A field study-design code comparison[J].Journal of Wind Engineering & Industrial Aerodynamics,2007,28(24):125～126
3. Shili Gong, Handbook for seismic design of buildings[M], Beijing, 2002
4. Ling Zhang,Minghua Zhao,Wei He. Working mechanism of two-directionreinforced composite foundation.J.Cent.South Univ.Technol,2007, 589-594
5. [5] Wakamastu K.Liquefaction case histories of voleanic deposits[A].Proe.lstInt. Conf.on Earthquake Geoteehnical Engineering[C].1995.19 — 23

The Effects of Packaging on Collision Energy Absorption of Automotive Seat Headrest

Chen Minggang[1, 2, a], Zhou Lei[3], Meng Zhenghua[1, b *],Wei Xuebing[3], Wang Hui[1]

[1]*Hubei Key Laboratory of Advanced Technology for Automotive Components,Wuhan University of Technology, Wuhan 430070, China.*

[2]*Hubei Collaborative Innovation Center for Automotive Components Technology,Wuhan University of Technology, Wuhan 430070, China.*

[3]*SAIC GM Wuling Automobile Co.,Ltd., Liuzhou, 545007, China.*

Abstract The automotive seat headrest plays an important role in the passenger protection during car crashes, and its structure parameters and performance have direct influence on the seat crashworthiness. In this paper, according to relevant regulations of GB11550-2009, collision simulation analyses of the seat headrest structure were carried out by Ls-dyna code. The law of different headrest packaging parameters on collision energy absorption was investigated to provide guidance for the headrest structure optimization design and improvement. The research results show that, with the increase of packaging of the headrest, the maximum acceleration and high acceleration duration time of head gradually decreases.

1 Introduction

In recent years, along with the increase of car ownership, car rear-end collisions have occurred occasionally. Much attention is paid to the safety of autos. The seat is an important component for ensuring occupant safety and comfort. The headrest is an important part of the seat and plays a key role in the car seat crash safety. During seat headrest design process, a large number of researchers have analyzed safety and comfort of car seat by the dynamic simulation method, especially at the beginning of the seat development stage[1-5]. For optimizing the structure design of seat headrest, a typical seat headrest structure was taken as analysis object and the laws of headrest packaging on collision energy absorption when the steel ball hits the headrest was investigated in this paper.

2 Model analysis

A typical seat headrest structure was chosen as the analysis object. According to relevant regulations of GB11550-2009, the steel ball's diameter is 165 mm and weight is 6.8 kg. At the top of the headrest down along the trunk line with the distance of 65 mm, the steel ball hits the headrest along the horizontal direction at the speed of V=22.4km/h to simulate head hits headrest process [6], as shown in Figure 1. The models of headrest frame, seat-back frame were meshed by high accuracy quadrilateral shell elements and headrest structure was meshed by hexahedral elements [7]. The number of shell elements is 35497 and the number of hexahedral elements is 65383.The material of steel ball is rigid body material (*Mat_Rigid). Headrest sponge material is recoverable, low density foam material with hysteresis effect (*Mat_Low_Density_Foam)[8].

Using Surface-to-Surface contact form simulate the crash between rigid ball and seat structure, as shown in Figure 2.

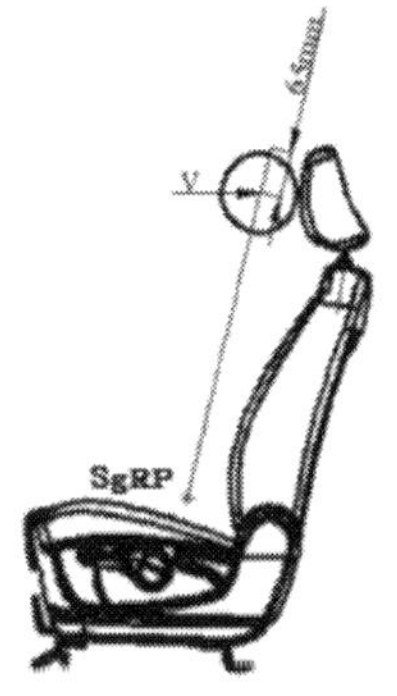

Figure 1. schematic diagram of collision between head and headrest

DOI: 10.1051/matecconf/20166501008

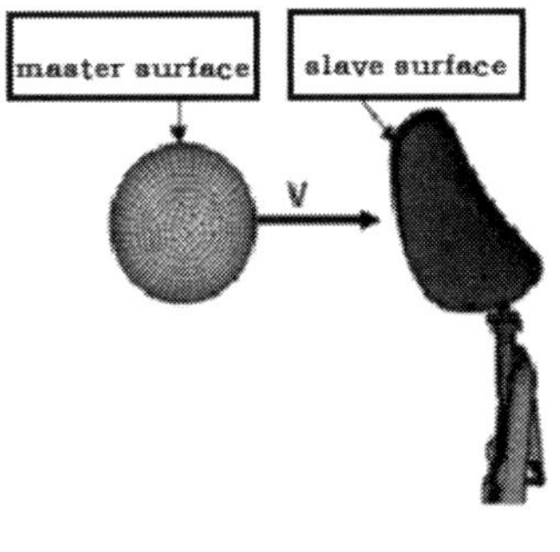

Figure 2. contact surfaces schematic diagram

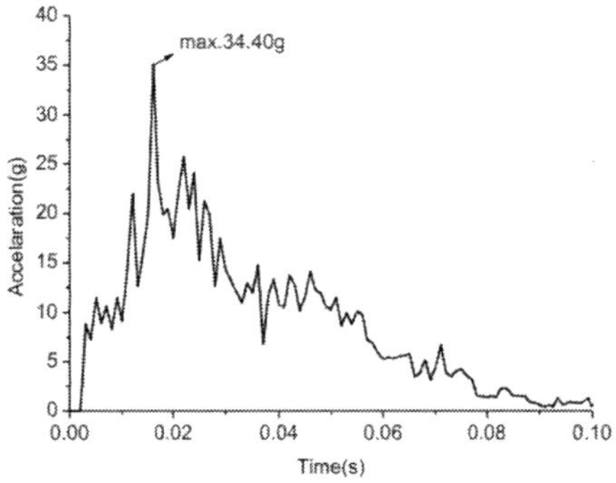

Figure 4. steel ball acceleration-time curve

3 Collision analysis and discussion

The main evaluation index for typical seat headrest crash safety in the paper are as follow: 1) the maximum acceleration of the steel ball when the steel ball hits the seat headrest; 2) the duration that the acceleration of the steel ball is more than 25g in the collision process (called high acceleration duration in this paper). The maximum acceleration and the duration of the high acceleration of the steel ball are smaller, the headrest's energy absorption and safety, comfort of the occupant are better.

3.1. Movement process analysis.

The movement process of the steel ball hitting the seat headrest and the acceleration curve of the steel ball are shown in figure 3, figure 4.During the collision, the ball at the speed of 6.69m/s hits the headrest. In the 2 ms, steel ball contacts with the headrest. Then the steel ball's acceleration gradually increases due to the headrest's buffering effect. With the steel ball continue to move, in the 17 ms, the acceleration reaches to the maximum 34.40g. With the decrease of the reaction force between the ball and headrest, the acceleration decreases gradually. The steel ball separates from the headrest at 91ms and the acceleration of steel ball reduces to about 0g and the speed tends to be stable. From the results showing in figure 3 and figure 4, the maximum acceleration of the steel ball hitting the headrest is 34.40g, which meets regulatory requirements.

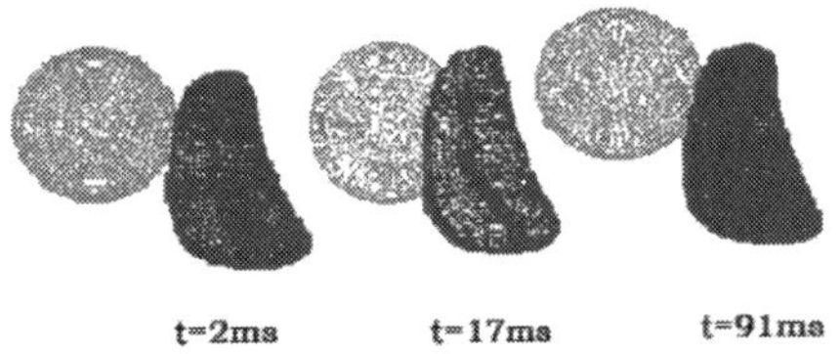

Figure3. steel ball movement diagram

3.2 The influence of headrest's packaging on collision energy absorption.

In the analyses, the headrest's packaging is changed by increasing the thickness of the headrest's both sides with the constant distance of L (shown as Figure 5).The distance difference B between headrest impact point and headrest both sides thickness represents the headrest's packaging.

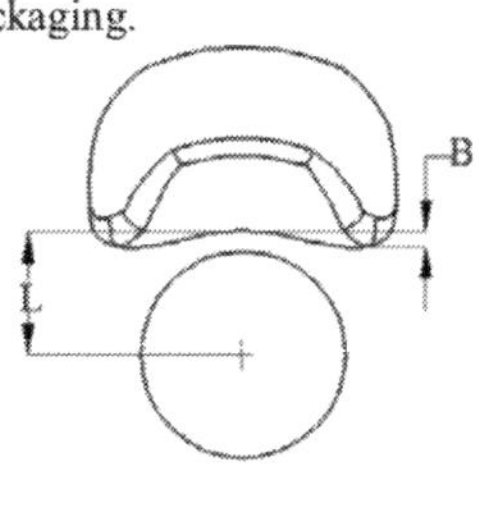

Figure 5. schematic diagram of headrest packaging

Under the condition of the headrest's thickness is 28mm, density is 60kg/m^{3}, the headrests which the packaging value B is 6 mm, 9 mm, 12 mm respectively are used for collision analysis. The steel ball acceleration curves are shown in figure 6. With the change of headrest's packaging, the maximum acceleration and high acceleration duration curves are shown in figure 7. From the figure 7, it can be found out, the maximum acceleration and high acceleration duration decreases with the increase of the headrest's packaging. This is because contact area between the steel ball and the headrest increases. Per unit area of the steel ball bearing reaction force decreases gradually. The increasing of the headrest's packaging improves the head's lateral stability significantly which can reduce additional head injury during collision [9].

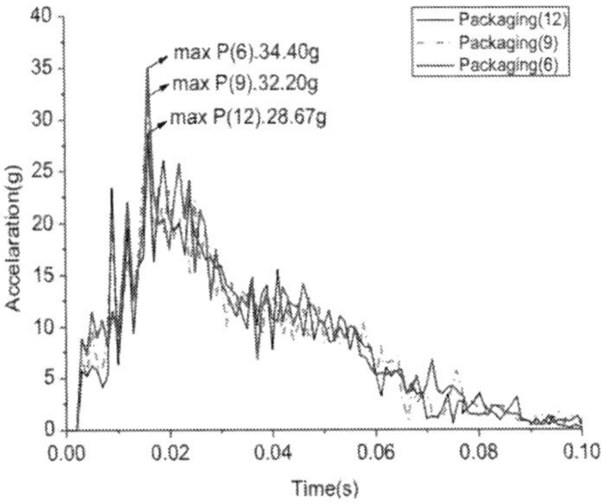

Figure6 steel ball's acceleration curves under the different packaging

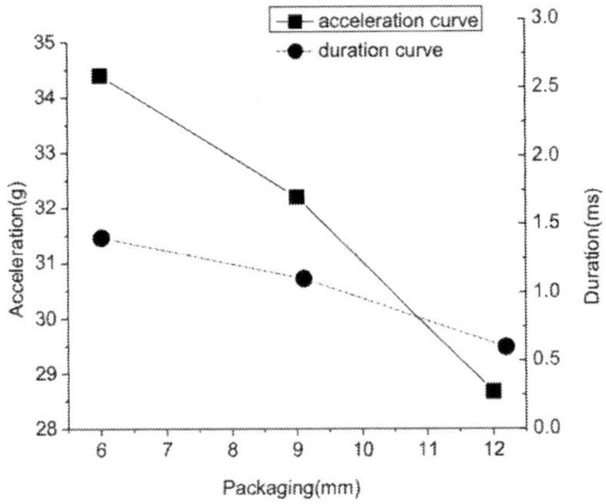

Figure 7 maximum acceleration and high acceleration duration curves

4 Summary

In this paper, energy absorption analysis of typical structure's headrest proves that the seat headrest meets regulatory requirement. Besides, the results show that with the increase of the headrest's packaging, head's the maximum acceleration and high acceleration duration of the head decrease gradually, which improve the headrest's energy absorption effectively.

Acknowledgements

This work was supported by the Natural Science Foundation of Hubei Province of China (No.2014CFB176) and Wuhan Planning Project of Science and Technology (No. 2013011803010606-4).

References

1. Brix, C. and Tok, C, Robust Design in Occupant Safety Simulation, SAE Int. J. Trans. Safety. 1(2), (2013), 241-260.
2. Fan Pingqing, Song Xinping, Zhao Bo, The analysis on impact capability of seat headrest for automobile, Machinery Design & Manufacture. 3 (2011), 118-120.
3. Lin Yi, Yao Weimin, Sun Dandan, Study on Automobile Seat Safety Performance During Impact, Journal of Beijing Institute of Technology. 25(1), (2005), 18-20.
4. Jin Jingxu, Song Xuewei, Huang Xuan, Cui Taisong, Effects of combinations of car seat headrest and recliner parameters on whiplash injury, Journal of Harbin Engineering University. 4(10), (2013), 1280-1286.
5. Cao Libo, Li Zhengkun, Gong Yongjian, A Study on the Development and Simulation on an Active Head Restraint, Automotive Engineering. 36(6), (2014), 746-750.
6. GB11550-2009 Strength requirement and test of automobile seats head restraints, Bei Jing Standards Press of China, 2009.
7. Zhang Junyuan, Huang Xuan, Tian Disi, Wang Hongbin, CAE Analysis and Design Modification for Rear Seat Safety During Vehicle Frontal Crash, Automotive Engineering. 33(9), (2011), 767-771.
8. Yao Weimin, Sun Dandan, Parameterization Design of Structure of Automotive Seat Headrest, Automobile Technology. 10 (2004), 4-7.
9. Zou Bowei, The Design of Comfort of Automobile Seat Based on Ergonomics, Automobile Technology. 4 (2009), 15-17.

MATEC Web of Conferences **65**, 01009 (2016) DOI: 10.1051/matecconf/20166501009
ICNSCM 2016

Non-linear Static Analysis of Offshore Steep Wave Riser

Hongdong Qiao[1,a], Weidong Ruan[1], Zhaohui Shang[1], Yong Bai[1]

[1]*College of Civil Engineering and Architecture, Zhejiang University, Hangzhou, China*

Abstract: A new solution combining finite difference method and shooting method is developed to analyze the behavior of steep wave riser suffering from current loading. Based on the large deformation beam theory and mechanics equilibrium principle, a set of non-linear ordinary differential equations describing the motion of the steep wave riser are obtained. Then, finite difference method and shooting method are adopted and combined to solve the ordinary differential equations with zero moment boundary conditions at both the seabed end and surface end of the steep wave riser. The resulting non-linear finite difference formulations can be solved effectively by Newton-Raphson method. To improve iterative efficiency, shooting method is also employed to obtain the initial value for Newton-Raphson method. Results are compared with that of FEM by OrcaFlex, to verify the accuracy and reliability of the numerical method.

1. Introduction

To overcome harsh environments and high pressure, six main configurations of steel catenary riser have been developed, such as free hanging catenary, lazy wave, steep wave, lazy S, steep S and pliant wave. Each configuration has its advantage to make itself better appropriate for particular applications. Among these configurations, steep wave riser is considered better to accommodate the geometry in deep water, and it is suitable for applications where the riser terminates at a subsea base or completion. As shown in Figure 1, typical steep wave riser is composed of a taut lower pipe connected to a positively buoyant section of catenary which itself joins the sagging part of the riser which in turn ends at the vessel [1]. It is similar to the lazy wave configuration, but has a nearly vertical connection at the seabed where a subsea base controls the touch-down point. Steep wave riser could accommodate large vessel offsets and effectively reduce the potential compression buckling at TDP. However, it requires a subsea base and subsea bend stiffener, and yet is able to maintain its configuration even if the riser fluid density changes.

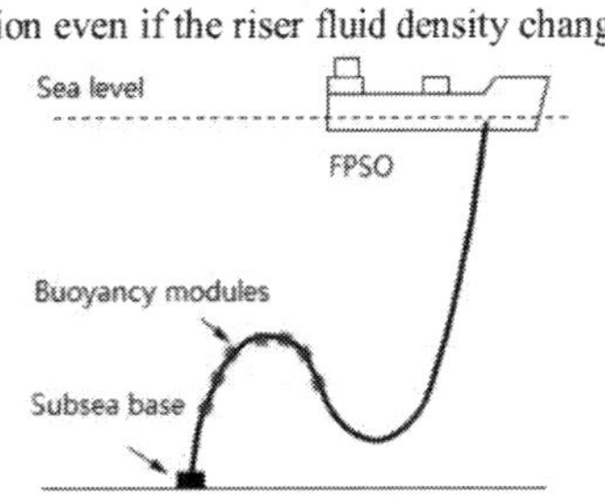

Figure 1. Configuration of steep wave riser

Research on steep wave riser is relatively few. Pinto and Lima [2] presented the final results of FSO steep-wave riser configuration design in the deepwater (815m) Roncador field and gave a complete report of that project. Then, numerical simulations by Riflex were performed to study the influence of the armour on the response of the steep-wave riser in 300 m water depth under irregular waves [3]. Sun and Bo [4] adopted the lumped mass method to perform the global analysis on the steep wave flexible riser, and found that the steep wave riser is not very sensitive to the ocean current. Steep wave configuration and lazy wave configuration used on the Auger TLP were specifically developed for application with a tanker FPSO located in a harsh environment [5]. And key mechanical design issues and material selection for different service conditions are discussed and budgetary costs for a range of typical field application were also presented in this paper. Santillan S.T. [6] put forward an elastic approach to describe the analysis of steep-wave flexible risers by a finite difference technique, and a free-vibration analysis was also conducted for small-amplitude oscillations.

This paper presents a numerical method to study the static two-dimensional behavior of steep wave riser by taking the effect of current loading into account. Due to the lifting effect of buoyancy modules attached to part of the riser, the behavior of the steep wave riser is subjected to large deformation and behaves great nonlinearity. Therefore, based on the large deformation beam theory and mechanics equilibrium principle, a set of non-linear ordinary differential equations describing the motion of the steep wave riser are developed. Then, finite difference method and shooting method are combined effectively to solve the ordinary differential equations with zero moment boundary conditions at both the seabed end and surface end of the steep wave riser. The resulting non-linear finite difference formulations can be solved by the Newton-Raphson method with the Jacobian matrix evaluated in an approximate manner. To verify the accuracy of the proposed method, FEM by OrcaFlex is employed and the results from the two methods show excellent consistency.

2. Configuration calculation

The behavior of the steep wave riser is described in terms of the position of the centerline of the riser according to the classical theory of rods [7]. Due to the attachment of buoyancy modules, the configuration of steep wave riser can be divided into three segments, as shown in Figure 2:

(1)Touch-down segment, it is the riser from the touch-down point (TDP) to the decline point (DP);

(2)Buoyancy segment, it is the riser from the decline point (DP) to the lift point (LP);

(3)Hang-off segment, it is the riser from the lift point (LP) to the hang-off point (HOP) at the vessel or platform.

In this figure, one global coordinate system (x, y) is established in the steep wave riser model by locating the origin at the touch-down point (TDP). And ocean current is set to be positive when flowing in the positive x-direction.

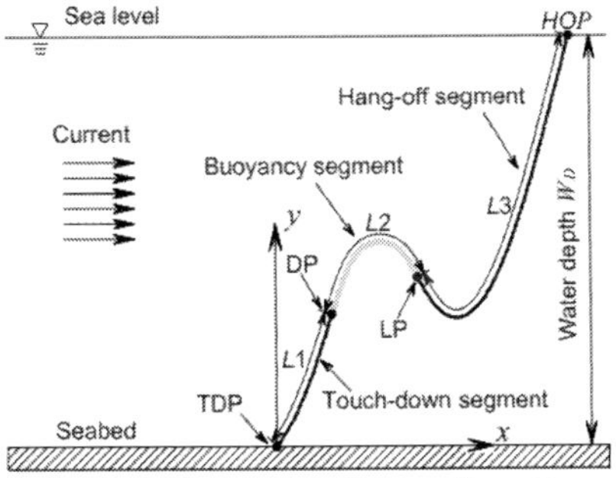

Figure 2. Steep wave riser configuration

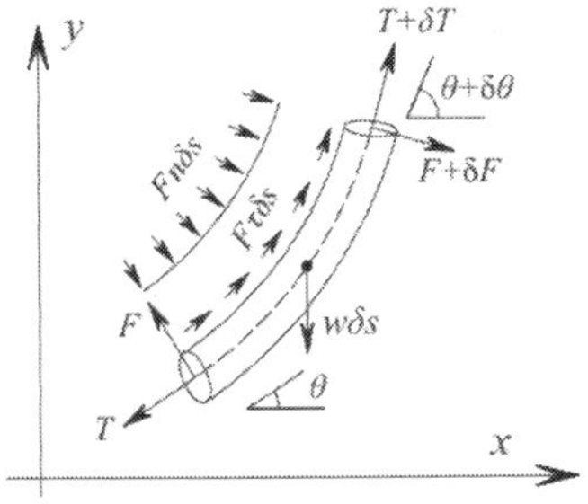

Figure 3. Large–angle deflections–forces acting on the touch-down segment

2.1 Touch-down segment

The arc length of the touch-down segment is L1. It is connected vertically to the seabed foundation by a subsea base, at the touch-down point. Due to the lifting force by the buoyancy segment and the displacement constraint of the subsea base, the configuration of the touch-down segment behaves like part of a simple catenary riser. Thus it can be seen that the behavior of the touch-down segment is subject to large–angle deflections. Based on the large deformation beam theory, the basic differential equations for the touch-down segment can be obtained from the force system acting on a differential segment of the touch-down segment, as illustrated in Figure 3 [8].

Resolving forces normal to the differential segment axis leads to

$$(F+\delta F)\cos\delta\theta-(T+\delta T)\sin\delta\theta-F+w\delta s\cos\theta+F_n\delta s=0 \quad (1)$$

$$F_n=\frac{1}{2}\rho_w C_d D\,ds\left|v_c\sin\theta\right|v_c\sin\theta \quad (2)$$

Where F and T represent shear force and axial tension, respectively; δF and δT represent the increment of shear force and axial tension; θ is the inclination angle of the differential segment from horizontal direction to axial direction; δθ represents the increment of the inclination angle; w is the submerged weight per unit length of the riser; δs is the length of the differential segment; ρw is sea density; vc is current velocity; D is the outer diameter of the riser; Fn represents the drag force in the normal direction per unit length of the riser, acting perpendicular to the riser axis. The drag force in the normal direction Fn can be obtained by Morison's equation.

As the length of the differential segment δs is very small, the increment variable δθ could be infinitely small. Therefore, cos(δθ)≈1 and sin(δθ)≈δθ. For simplification, Equation (1) then can lead to

$$\frac{dF}{ds}-T\frac{d\theta}{ds}+w\cos\theta+F_n=0 \quad (3)$$

Furthermore, the shear force F herein is related to the bending moment by F=dM/ds. Hence, Equation (3) can be derived as following:

$$\frac{d^2M}{ds^2}-T\frac{d\theta}{ds}+w\cos\theta+F_n=0 \quad (4)$$

Based on the large deflection beam theory, the bending moment then can be obtained:

$$M=EI\kappa=EI\frac{d\theta}{ds} \quad (5)$$

Where, EI denotes the flexural stiffness of the riser and κ denotes the curvature of the riser.

Combining Equation (4) and Equation (5), the final normal mechanics equilibrium differential equation can be expressed as:

$$EI\frac{d^3\theta}{ds^3}-T\frac{d\theta}{ds}+w\cos\theta+F_n=0 \quad (6)$$

On the other hand, resolving forces parallel to the segment axis leads to

$$(F+\delta F)\sin\delta\theta+(T+\delta T)\cos\delta\theta-T-w\delta s\sin\theta+F_\tau\delta s=0 \quad (7)$$

$$F_\tau=\frac{1}{2}\rho_w C_\tau \pi D\,ds\left|v_c\cos\theta\right|v_c\cos\theta \quad (8)$$

Where, Fτ represents the drag force in the axial direction per unit length of the riser, and Cτ is the drag coefficient in the axial direction.

For simplification, Equation (7) then can be simplified as:

$$F\frac{d\theta}{ds}+\frac{dT}{ds}-w\sin\theta+F_\tau=0 \quad (9)$$

Then, the final axial mechanics equilibrium differential equation can be expressed as:

$$EI\frac{d^2\theta}{ds^2}\frac{d\theta}{ds}+\frac{dT}{ds}-w\sin\theta+F_\tau=0 \quad (10)$$

As both the axial deformation and shear deformation are not considered in the analysis, the horizontal and vertical displacements in the global coordinate (x, y) then can be expressed:

$$x=\int\cos\theta ds\ ,\quad y=\int\sin\theta ds \quad (11)$$

Once the displacements, inclination angle and curvature along the riser are obtained, the bending moment M and shear force F can be derived as follows:

$$M=EI\kappa=EI\frac{d\theta}{ds} \quad (12)$$

$$F=\frac{dM}{ds}=EI\frac{d^2\theta}{ds^2} \quad (13)$$

It should be noted that shear force is set to be positive when in clockwise direction.

2.2 Buoyancy segment

The arc length of the buoyancy segment is L2. To decrease the upper end tension required at the vessel and decouple the vessel motions from the touch-down point of the riser, buoyancy modules made of syntactic foam or thin-walled air can are uniformly attached along the long length of the buoyancy segment. As the buoyancy force of the buoyancy modules and riser in the buoyancy segment is larger than their weights, the configuration behaves like an arch bend. To simplify the numerical analysis, a uniformly distributed lifting force is acting on the buoyancy segment, which is equivalent to the loading combination of the buoyancy force and weight. The uniformly distributed lifting force here is called "equivalent submerged weight unit length". Besides, the equivalent outer diameter and equivalent flexural stiffness are considered in the buoyancy segment. The adoption of the equivalent outer diameter here is to make sure that the buoyancy force of the equivalent buoyancy segment is equal to the buoyancy combination of the buoyancy modules and riser unit length. And the equivalent flexural stiffness is considered, due to some effect of these buoyancy modules on the flexural stiffness of the buoyancy segment. However, the equivalent flexural stiffness of the buoyancy segment is generally close to the flexural stiffness of the riser due to the small length of buoyancy modules attached. Thus it is generally acknowledged that the equivalent flexural stiffness of the buoyancy segment is equal to the flexural stiffness of the riser. The vertical displacement of the highest point in the arch bend is herein set as yarch. Due to its large-deflection behavior, the configuration analysis of the buoyancy segment is also based on the large deflection beam theory. According to the mechanics equilibrium principle, the governing differential equations for the buoyancy segment are similar with those of the touch-down segment. The main difference is that the flexural stiffness EI, submerged weight w and outer diameter D in Equations (1)-(13) should be replaced by the equivalent values EIe, we and De, respectively.

2.3 Hang-off segment

The arc length of the hang-off segment is L3. It is hung off by a ball joint to prevent high bending moment induced by wave, current and vessel motion, at the hang-off location. However, as the study is confined to the two-dimensional static analysis, the effects of wave and vessel motions are not considered in this study. Due to the lifting effect of the buoyancy segment and the vessel, the configuration of the hang-off segment behaves like a sag bend. The vertical displacement of the lowest point in the sag bend herein is set as ysag. As the mechanical behavior of the hang-off segment is precisely the same with that of the touch-down segment, the governing differential equations for the hang-off segment are also the same with that of the touch-down segment.

2.4 Boundary conditions

The touch-down location of the steep wave riser is generally connected to a subsea base, controlling the touch-down point. Therefore, the boundary condition of the touch-down point in the global coordinate system (x, y) can be expressed as:

$$M(0)=0 \quad (14)$$

On the other hand, the hang-off location of the steep wave riser is usually hinged to the vessel or platform by a ball joint. And the inclination angle of the hang-off point can be measured by instrument. Thus, the boundary condition of the hang-off point in the global coordinate system (x, y) can be expressed as:

$$y(L_1+L_2+L_3)=W_D \quad (15)$$

$$\theta(L_1+L_2+L_3)=\theta \quad (16)$$

$$M(L_1+L_2+L_3)=0 \quad (17)$$

Where, WD represents water depth.

3. Numerical solution

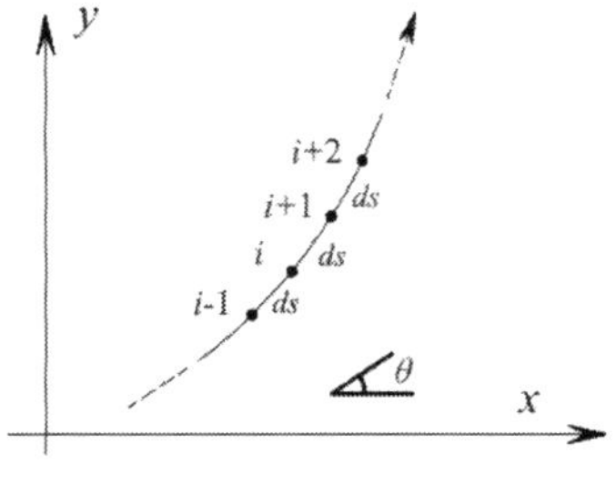

Figure 4. Forces acting on the touch-down segment

In this study, the behavior of the steep wave riser is mainly dominated by Equation (6) and Equation (10). According to the two differential equilibrium equations and boundary conditions, it can be found that the essence of steep wave riser's global analysis is a multipoint boundary value problem. Thus the finite difference

 DOI: 10.1051/matecconf/20166501009

method is applied to solve the two differential equations numerically in terms of the coordinates of adjacent nodes. For the case of equally spaced nodes, it can be expressed at node i, in terms of the element length ds between nodes and the angles at four nodes: the node itself, the two nodes immediately above it, and the one node immediately below it, as shown in Figure 4. The node number here is marked from 1 to (n+1), where the touch-down point is set as node 1 and the hang-off point is set as node (n+1). By considering the effect of ocean current, equations (6) and (10) can be expressed numerically in finite difference form by equations (18) and (19) respectively:

$$EI\frac{\theta_{i+2}-3\theta_{i+1}+3\theta_i-\theta_{i-1}}{(ds)^3}-T_i\frac{\theta_{i+1}-\theta_i}{ds}+w\cos\theta_i+\frac{1}{2}\rho_w C_d D ds|v_c\sin\theta_i|v_c\sin\theta_i=0 \quad (i=1\sim n) \tag{18}$$

$$EI\frac{\theta_{i+1}-2\theta_i+\theta_{i-1}}{(ds)^2}\cdot\frac{\theta_{i+1}-\theta_i}{ds}+\frac{T_{i+1}-T_i}{ds}-w\sin\theta_i+\frac{1}{2}\rho_w C_\tau \pi D ds|v_c\cos\theta_i|v_c\cos\theta_i=0 \quad (i=1\sim n) \tag{19}$$

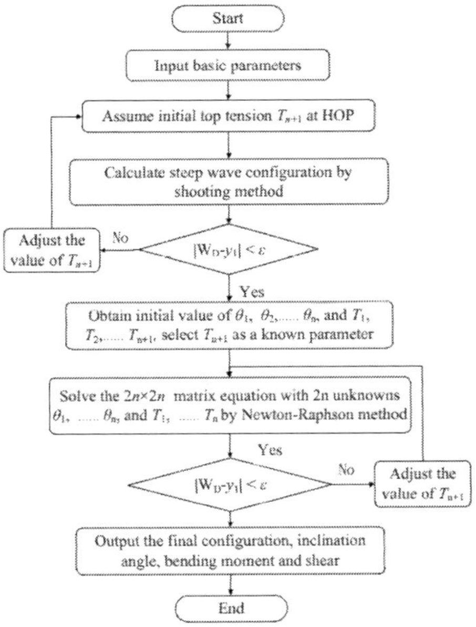

Figure 5. Flow chart of numerical calculation

When the length of the riser reaches between L1 and (L1+L2), that is to say when calculating the behavior of boundary segment, the flexural stiffness EI, submerged weight w and outer diameter D should be replaced by the equivalent values EIe, we and De, respectively. Then a 2n×2n matrix equation can be formulated containing n+3 angle unknowns (θ0, θ1, θ2,⋯⋯,θn+1 and θn+2) and n+1 tension unknowns (T1, T2, ⋯ ⋯ ,Tn and Tn+1). According to the boundary condition (14) and (17), the value of θn+2 can be expressed in terms of θn+1 and θn, and θ0 in terms of θ1 and θ2. Besides, the value of θn+1 can be obtained from the boundary condition (16). Finally, it can be found that there are still 2n+1 unknowns (θ1, θ2,⋯⋯,θn and T1, T2,⋯⋯, Tn+1) left in the matrix equation. The above problem can be effectively solved by shooting method. In addition, to improve the efficiency of Newton-Raphson iteration method, the initial assumed parameters used in this iteration method could be derived by shooting method [9]. Figure 5 shows the main procedures of the numerical solution. In the flow chart, ε respects the specified small quantity.

4. Comparison and discussion

In order to verify the accuracy and reliability of the numerical method, comparisons are performed with the results obtained from the proposed method and FEM by OrcaFlex. To perform the analysis, basic properties of the riser and environment are shown in Table 1 and Table 2, respectively. Comparisons of steep wave configuration, tension, bending moment and shear force are shown in Figures 6-9, respectively. From these figures, it can be seen that the results obtained from the two methods coincide well. It can strongly verify the accuracy and reliability of the proposed method. Differences between the numerical method and FEM by OrcaFlex are shown in Table 3

Table 1. Riser parameters

Items	Value
L_1(m)	450
L_2(m)	672
L_3(m)	1076
w(N/m)	501.7
w_e (N/m)	-785.9
D(m)	-0.220
D_e(m)	-0.460
EI(KN·m^2)	4.736E2

(Note: the flexural stiffness of the boundary segment is assumed to be the same with the other two segments, that is to say EI=EIe.)

Table 2. Environmental parameters

Items	Value
W_D(m)	1500
ρ (kg/m^3)	1025
v_c (m/s)	0.2
θ (deg)	80
C_d	0.7
C_τ	0.008

DOI: 10.1051/matecconf/20166501009

Table 3. Results of numerical method VS FEM by OrcaFlex

Items	Numerical method	FEM by OrcaFlex	Difference (%)
x_{HOP} (m)	1018	1012	0.59
y_{arch} (m)	891.6	893.2	1.79
y_{sag} (m)	755.1	755.1	0.01
T_{HOP} (KN)	454.7	454.7	0.00
T_{TDP} (KN)	225.9	226.4	0.22
M_{arch} (KN·m)	-4.520	-4.588	1.48
M_{sag} (KN·m)	2.919	2.939	0.68
F_{DP} (KN)	-0.1145	-0.1187	3.54
F_{LP} (KN)	0.8261	0.8258	0.04

As shown in Figure 6, the steep wave configurations from the two methods seem in perfect consistency. It can be seen that the difference mainly happens at the hang-off segment. The horizontal distances from the hang-off point to the touch-down point obtained from numerical method and FEM by OrcaFlex are 1018m and 1012m respectively, and the difference is less than 3.0%. Besides, the highest vertical locations of the arch bend from the two methods are 891.6m and 893.2m, respectively. And the lowest vertical locations of the sag bend from the two methods are 755.1m and 755.1m, respectively.

The tensions illustrated in Figure 7 seem also in excellent agreement. It can be found that three local maximum tensions occur at the hang-off point, lift point and decline point, respectively. Among the three local maximum tensions, the tension at the hang-off point seems to be the maximum, arriving at 454.70KN. While under the same condition, the upper end tension in simple free hanging configuration without the addition of buoyancy modules can reach up to 915.63KN, which is greater than that in the steep wave configuration. It can be seen that steep wave configuration can reduce the upper end tension effectively.

Figure 8 shows the comparison of riser's bending moments, and the overall bending moments almost perfectly match. As expected, two local maximum bending moments occur at the peak point in the arch bend and the lowest point in the sag bend. In addition, the values of bending moments at the LP and DP turn out to be zero. And it can be found that the bending moments in the neighborhood of the LP and DP change very obviously. The maximum bending moments in the steep wave configuration from numerical method and FEM by OrcaFlex are -4.520KN·m and -4.588KN·m respectively, which are very close.

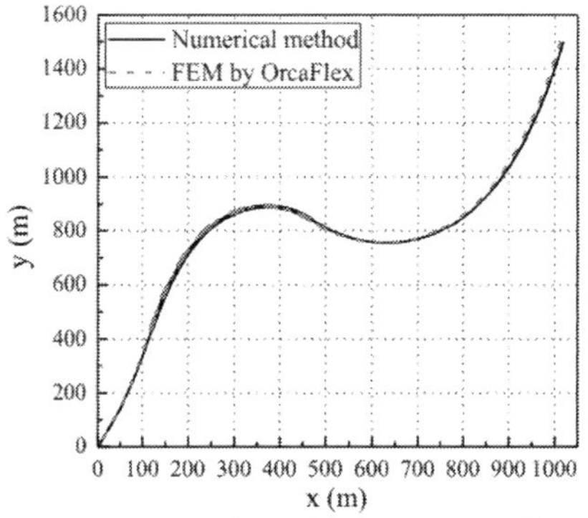

Figure 6. Comparison of steep-wave configurations

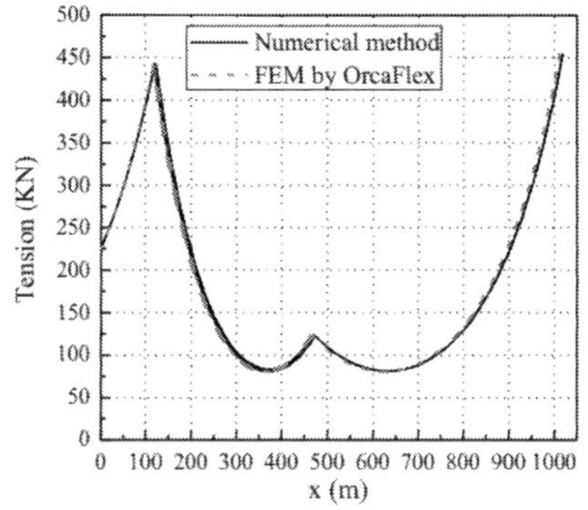

Figure 7. Comparison of steep-wave tensions

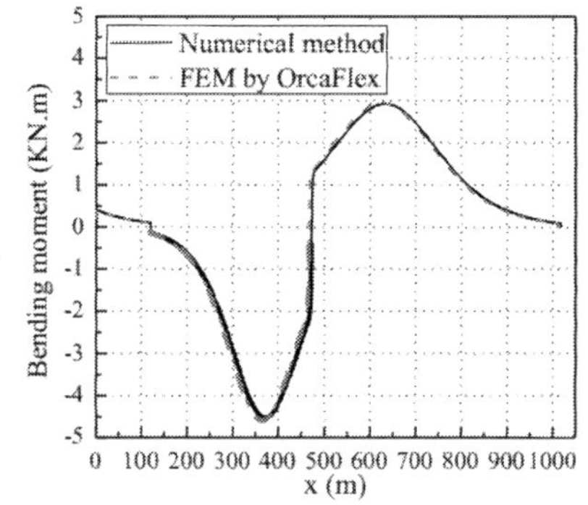

Figure 8. Comparison of steep-wave bending moments

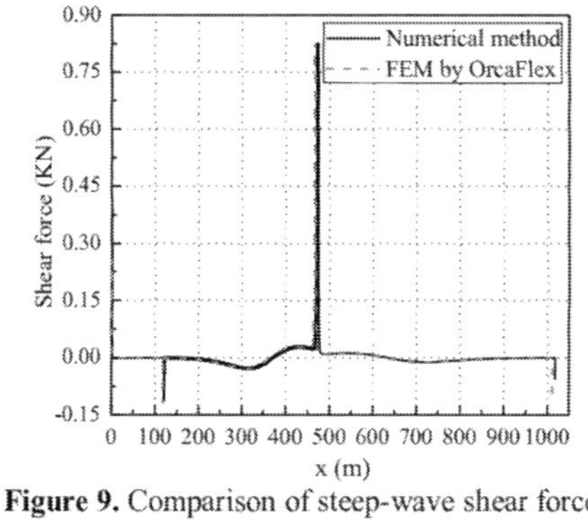

Figure 9. Comparison of steep-wave shear forces

The comparison of riser's shear forces is shown in Figure 9. The overall shear forces in Figure 9 match very well, except at the two locations of HOP and TDP. Furthermore, two local maximum shear forces occur at the DP and LP and the shear force in other location is very close to zero. The local maximum shear forces in the neighborhood of DP from numerical method and FEM by OrcaFlex are -0.1145KN and -0.1187KN respectively. And the local maximum shear forces in the neighborhood of DP from numerical method and FEM by OrcaFlex are 0.8261KN and 0.8258KN respectively.

5 Conclusions

This study presents a reasonable numerical method to study the mechanical behavior of planar steep wave riser considering the effect of ocean current. According to mechanics equilibrium principle, two non-linear ordinary differential equations are set to for each differentiation element of steep wave riser. Finite difference method and shooting method are combined to solve the ordinary differential equations and the resulting non-linear finite difference formulations can be solved by the Newton-Raphson method. The accuracy and reliability of the proposed method has been well verified by comparing FEM by OrcaFlex. In conclusion, the comparison analyses have demonstrated the ability of the proposed method in predicting the static behavior of steep wave riser. Further research needs to be carried out on dynamic response of steep wave riser.

References

1. F.B. Seyed, M.H.Patel. Mari. Stru. 5(2),121(1992)
2. P. Pinto, G.Lima. 20th Int. Conf. on Offs. Mech. and Arct. Engi. 133(2001).
3. B. Bonnemaire, 14th Inte. Offs. and Pola Engi. Conf. 236(2004).
4. L. Sun, B. Qi, Mari. Sci. and Appl. 10(4): 478(2011)
5. S. A. Hatton. FPSO Worl. Congr. & Tech. Exhi, 800(1995).
6. S.T. Santillan, L.N. Virgin, R.H Plaut R.H, J Offs. Mech Arct. 132(2010)
7. D.L. Garrett.. J. Ener. Reso. Tech, 104(1982)
8. C. P.Sparks. Fund. of mari. Rise. Mech: basi. Prin. and simp.anal, (2007).
9. Y.Bai, B.Yu, P.Cheng, The 22nd Intl.Offs.and Pola. Engin. Conf, **2** 103(2012)

MATEC Web of Conferences **65**, 01010 (2016) DOI: 10.1051/matecconf/20166501010
ICNSCM 2016

Research on patent portfolio design by Using of TRIZ Method

Li Hui[1], Liu Luguang[2], Huo Jiangtao[3], Xu Bo[4]

National technological Innovation Method and Tool Engineering Research Center, Hebei University of Technology, China

Abstract: By integrating patent portfolio design ideas and TRIZ innovative approaches, a patent portfolio design model is proposed in this paper, which is based on the enterprise's core technology. Basing on this model, the TRIZ based portfolio design process is developed via matching the TRIZ innovative methods and the types of patent portfolio and constructing patent portfolios using homologous TRIZ innovative methods. Taking TRIZ as the basic theory and synthesizing innovative methodologies, patent jurisprudence, and management, interdisciplinary research of collaborative innovative design is conducted.

1. Introduction

Patent portfolio is strategic patent collection, which will consolidate the patent owner's patent competitive advantage through combination of patents that are obvious differences and closely related. Patent portfolio design is the innovative design behavior of resources allocation, technology excavation and patent portfolio, which is patent-portfolio formation oriented.

Portfolio design ideas originate from economic management field. In 1952, Ma Horowitz (Markowitz) [1] an American economist, firstly proposed the portfolio management theory. In early 1990s, Brockhof [2] introduced this theory to the field of corporate patent strategy in his pioneering patent portfolio model. Current researches on patent portfolio mostly focus on the area of management. Sccot [3] provides a mechanism for patent portfolio analysis and decision-making to help companies effectively implement the developing strategy of the R & D patent and enhance their research and development capabilities. In Ziedonis' research [4], companies' core technology was found by searching the scale and number of the best patent portfolios. Then the peripheral technology system around the core technology system was built to develop the patent portfolios in the field of core technology. Grindley [5] studied the combination of patent licensing and cross-licensing and hold the view that the patent portfolio model is helpful for the choice of internal and external R & D. Kang Yuhang [6] take the multi-scaling MDS as a patent portfolio analysis tool to demonstrates the combination relations between patent data item sets in the form of visualization for business analysis. In this paper, various problem solving and analyzing tools are applied in patent portfolio design processes, such as patent analysis, patent excavation, patent applications and ultimately the patent portfolio.

2. TRIZ Theory and patent portfolio design

DOI: 10.1051/matecconf/20166501010

2.1 Portfolio design model.

Patent portfolio design is an innovation design method based on patent portfolio guidance. Its types and graphical representation are as shown in Tab.1:

Table 1. The type of portfolio design model

Name of patent portfolio	**Legend said**
Cluster patent portfolio	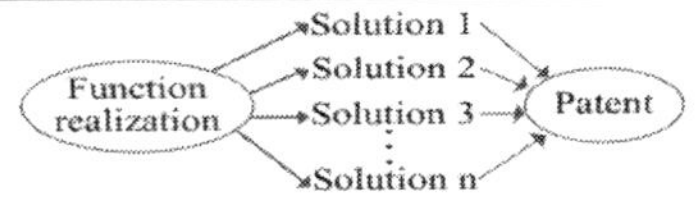
Parachute patent portfolio	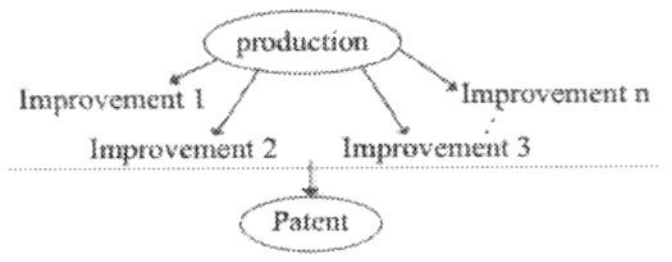
Galaxy type patent portfolio	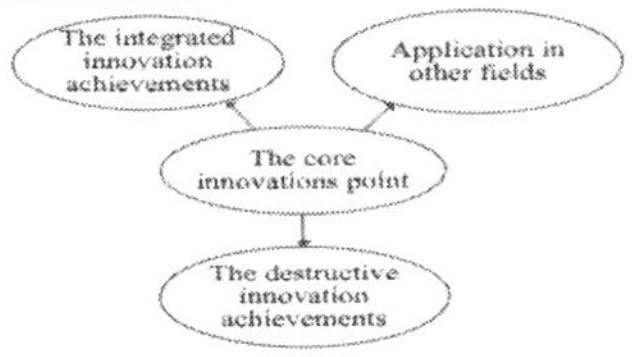
Chain patent portfolio	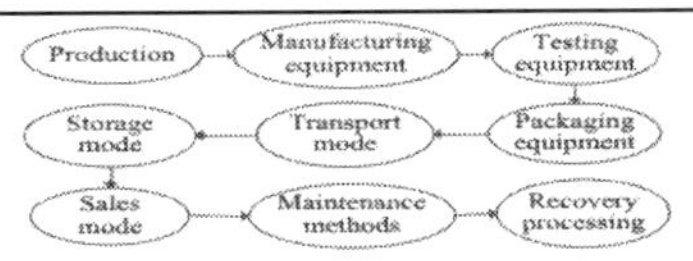

Comprehensive above several kinds of portfolio design models, based on the core technologies in enterprises, this paper proposes a new patent portfolio design models, as shown in Fig.1.

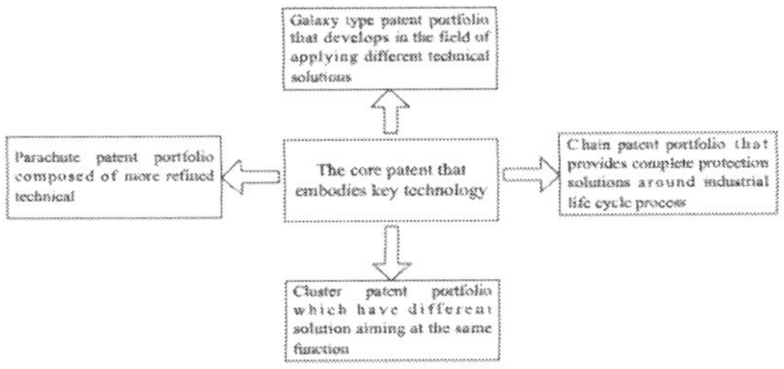

Fig.1 Patent portfolio design models based on the core technologies in enterprises

2.2 TRIZ-based patent portfolio design process.

Patent portfolio design process based on TRIZ theory consists of five steps, as shown in Fig.2.

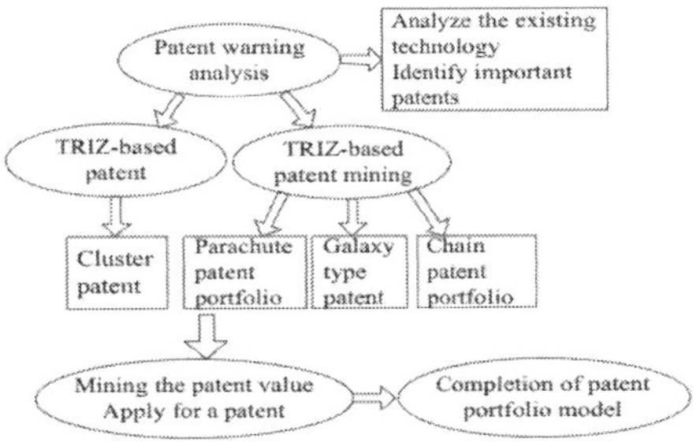

Figure 2: Portfolio design process based on TRIZ

2.2.1 Patents warning analysis.

Patent early warning analysis based on TRIZ theory mainly includes three parts content. First, functional analysis based on TRIZ is used for multi-dimensional technology decomposition in the target technology. Second, the target technology based on the decomposition is used to search for each function or function points. The third step is to build patent analysis tree based on the strategic integration of functional analysis and patent portfolios.

2.2.1.1 Target technology of multidimensional angle-resolved based on functional analysis.

Functional analysis is an important theoretical and analytical tool in TRIZ. The purpose of existence of product is its function. By establishing a functional model, the functional analysis of the product analyzes all of the functions which require implementing a function, until the analysis to the smallest functional unit that is the functional element. Taking the function can reveal function and meaning of functional components, parts or assembly unit into consideration, in function block diagram, it can also mark the role of the relationship between the components, create an object from the input to the output of the model of complete product structure graphical analysis. On this basis, function can classify, combine and decompose the

DOI: 10.1051/matecconf/20166501010

target technology so that it can achieve the multidimensional decomposition of the target technology.

2.2.1.2Search for each function or function points based on the decomposition of the target technology.

Only a detailed analysis of target technique can ensure the comprehensiveness and accuracy of the data. A list of information on functional analysis of the target technology and the functions of each points or functional element covered by equivalent characteristics are to construct Boolean logic retrieval type, then retrieval the technical solutions which are related to technological problems of the functions of each points or functional element in the existing technology, read retrieved patent documents, identify the important patents and patent portfolio and establish technical efficiency corresponding figure as a form of subsequent design technical constraints.

2.2.1.3The construction of patent analysis tree.

By retrieving the target technology, and identify the most relevant patents, that the important goal of patents, and then be expanded around the patent search, collect important patents that already exist in the prior technology combination from the four directions, build important patent parse tree, as shown in Fig.3, be important in the analysis tree Patent No. marked as inspired subsequent research and reference, to avoid infringement of the right combination of design, and the opportunity for each mining technology under patent portfolio.

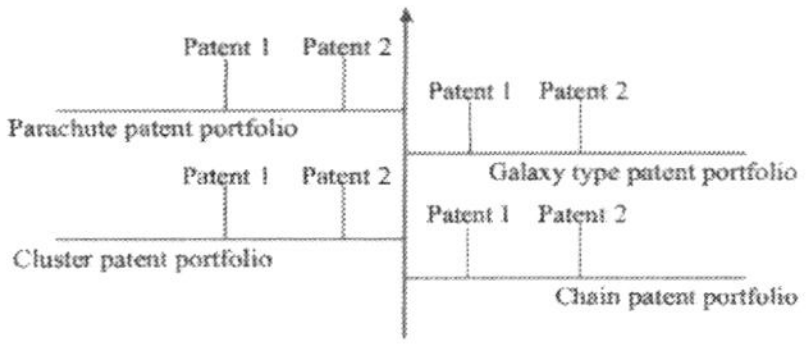

Fig.3 Patent tree

2.2.2TRIZ -based patent circumvention design.

Patent circumvent design is to bypass the protection scope of the rights of a certain patent and realize innovative designs of same function. A variety of design schemes after patent circumvented can constitute a cluster-type patent portfolio of same function and different means of accomplishing. Patented circumvention design based on functional analysis method in TRIZ theory. A patent circumvention designing process using TRIZ theory is given as follows. First, each branch function or functional element in the functional model established by information analysis is to be analyzed; then functional elements of interest are to be selected; Second, standard engineering parameters and corresponding invention principles are to be abstracted. 39 standard engineering parameters are given by conflict resolution theory of TRIZ, then, corresponding invention principles can be found out according to the characteristic parameters and the deterioration parameters to be improved. The same functional elements can be re-solved by means of the tips of invention principles. Third, re-solve the analogically analyzed functional elements, by means of the patents retrieved from expanding areas through standard engineering parameters or patent cases corresponding to the invention principles suggested by innovative software, a number of innovative ideas can be produced to re-solving the same functional elements. Fourth, combine the functional elements according to patent infringement law, develop a non-infringing technical proposal, and then patent design is to be achieved.

2.2.3Patented mining design based on TRIZ.

Using different methods in TRIZ, more patents can be excavated, and these will improve autologous competitiveness. And the design can be divided into the following situations.

2.2.3.1 Patent mining for parachute patent portfolio.

Patent mining facing to parachute patent portfolio improves existing core technology solution which mainly uses substance field model, conflict analysis and the invention principles in TRIZ theory, which includes following two application routes.

The first is substance -field analysis and its applications. On the basis of material field analysis of functional units, material field analysis is an important tool to analyze problems in TIRZ. The route uses all kinds of pre-set standard material field model to discover problems in existing function, then, take advantage of 76 standard solutions to excavate problem solution. The second is conflict analysis and application of the invention principles. On the basis of existing technology, core patents are done by conflict analysis, and technical and physics conflicts are found. These principles can be used to improve weakness in existing technology, and perfect solution is formed.

2.2.3.2 Patent mining for galaxy type patent portfolio.

Building galaxy type patent portfolio are mainly looking for expanding application of existing technology in other fields, this expansion includes three aspects.

One is that new product purpose in other areas, such as washing powder has been applied in the fattening farm fields, and it corresponds to apply for use patent. The second is products of integration innovative in other areas are developed. Thirdly, developed certain products which property reduces can satisfy the requirements of the special group. The invention of corresponding elements reduced may be applied.TRIZ methods which correspond to above three kinds of galaxy type patent portfolio, respectively are as follows. The first resource analysis method is using resource analysis to expand application fields, and is a method to screen and classify existing resource in TRIZ theory. Aiming at some universal patent innovation, it can screen all kinds of resource, and search for new fields of application. The second method using hybrid integrated innovation method, expands technological opportunity with valuable technical point and integrating with other products, and is aimed at excavating new products. Disruptive innovation is aiming at the existing products with mainstream technologies by using cutting tool in TRIZ method.

2.2.3.3 Chain patent mining for the whole product life cycle.

From the implementation perspective of a particular technology or product industrialization, chain patent mining for the whole product life cycle develops and mining patents aiming at corresponding supporting and supporting technology in each link of core product, such as production equipment to packaging, marketing, transportation and even repair and recovery methods, etc. They can not only protect core technical solution powerfully, but also determine the competitive status of enterprise. This method can mining patent existing industrial chain cycle, according to law of demand evolution of TRIZ tools.

2.2.4 Evaluation the value of patents and patent applications.

Patent application as the one of the most important aspects of the innovation for the protection of intellectual property, is very important in the design of enterprise patent portfolio, the purpose of the patent portfolio design is obtained a wide scope of protection after an application for a patent for innovative solutions, assess the value of patents and patent applications involves two main steps: The first is based on the TRIZ innovation level of innovation to sort and classify, and give the corresponding intellectual property protection strategy; The second is to control patent analysis tree, feasibility evaluation of innovative solutions to avoid stepping on other people's patents mines and build their own patent portfolio.

3. Conclusions

TRIZ theory is an advanced method of auxiliary enterprise innovation, enterprise innovation achievements in the protection of the process, building a reasonable scope of protection of the patent portfolio let the enterprise the core technology and key technology of a relatively wider scope of protection, the protection of a relatively longer time in order to ensure the effectiveness of innovations. The innovative design in the beginning, to import the patent portfolio theory, can reduce the blindness of enterprise intellectual property protection, to prevent a single patent protection range and the emergence of unreasonable patent portfolio, to win intellectual competition.

Acknowledgements

This work was supported by the Natural Science Foundation of China [project numbers: 51275153] and the Special Fund of Innovation Methods Work [project number: 2012IM040300].

References

[1] MARKOWITZ.H, J. The Journal of Finance, Portfolio selection, **1**, 77-91 (1952)
[2] BROCKHOFFK, J.Technovation,Instruments for patent data analysis in business firms, **12**, 41-58 (1992)
[3] Sccot. A R, J. The Licensing Journal, A computer-friendly microeconomic patent portfolio valuation algorithm, **12** ,14-18 (2001)
[4] Z iedon is R H, D. University o f California, Patent protection and firm strategy in the semi conductor industry, Berkeley, (2000)
[5] Grindley P C, Teece D J,J. California Management Review,Managing intellectual capital licensing and crossing- licensing in semiconductor and electronics, **2** , 8-41 (1997)
[6] Kang Yu-hang, J. Scientific Research, Patent portfolio and application-based mapping of multidimensional scaling, **1** ,30-35 (2009)

Artificial Neural Network Analysis of the Impact of Sample Output Accuracy

Yu Zhong Zhang[1,a], Shao Yun Song[1]

[1]*School of information Technology and Engineering,Yuxi Normal University, Yuxi,Yunnan,China*
[2] *School of information Technology and Engineering,Yuxi Normal University, Yuxi,Yunnan,China*

Abstract. The impact of artificial neural network model output precision technology widespread attention. Quality sample study of neural network output accuracy is not much affected, most of the research is the structure (number of layers and the number of nodes), the impact of this paper to analyze samples of artificial neural network output for the neural network, to improve the output of neural network accuracy is important.

1 Introduction

The neural network to prediction areas, there are many problems to be solved. One of the most prominent problem is there is no standard method to determine the most suitable one neural network structure. Since the parameters that affect neural network prediction capabilities are many. This paper most commonly used BP neural network, quality of the sample neural network prediction accuracy is analyzed in detail and research, and on this basis, to give specific examples of the samples after optimization.

2 Effects of sample quality on the network

Neural networks samples are divided into training sample and testing sample, the quality of the training samples to some extent determines the prediction accuracy. First, there is a big difference between the mean of training samples and the samples to be predicted, the prediction error will be increased as the long-term training. Secondly, the training error increases with training samples will be predicted and the difference between the sample mean increases. Again, with increasing training error would be predicted between the training sample and the sample variance difference increases [1]. Here is the literature [1] method of sample quality analysis.

The neural network prediction error in artificial [1] is $e = e_m + e_t + e_r$, where e is the prediction error, e_m is model error, It was established by the regression model and the differences caused by the actual system, e_t is the final training error, e_r is random errors in the artificial neural network training and prediction process . The existence of e_m and e_t are inevitable, and e_r is $e_m = e_f + e_d$, where e_f is the error between the actual output and predicted output values, which reflects the sample quality; e_d by embedding an incorrect dimension error caused it can be the appropriate number of input neurons by selecting eliminated.

In order to evaluate the quality of training samples, according to e_f proposed indicators "consistent degree". defines a pseudo-distance DCT_{P-D} in paper [1], but pseudo-distance calculation is quite complex and difficult to achieve for analysis and applications of sample quality. The following analysis of the impact on the prediction accuracy of the training sample covariance ratio statistic.

Assume training samples is θ, $\hat{\theta}$ is the output of neural networks, $\hat{\theta}_i$ is the neural network output excluded from eection θ of i -th data points . The ratio statistic covariance of the i-th excluded data points is $CR_i = \left|\hat{\text{cov}}(\hat{\theta})\right| / \left|\hat{\text{cov}}(\hat{\theta}_i)\right|$, It shows the effect on the accuracy of the neural network output removed after the i-th data point, From the accuracy portrayed the importance of the i-th data point. The value of $|CR_i - 1|$ greater then the impact of θ_i on the prediction accuracy (neural network output) greater. When using neural networks to predict PB samples were first screened for training, excluding the results of a small impact on the network output sample points.

Samples streamlined. Elementary row transformation matrix to maintain a linear relationship between the matrix column vector, this conclusion can be used to

[a] Corresponding author: zh1011@yxnu.net

streamline our sample data, sample data so streamlined to maintain a linear relationship between each attribute fields. Training the neural network is actually a given sample in real time to adjust the network connection weights through the process, the results of the sample pretreatment convergence for network training play a key role.

3 Analysis of practical examples

In this paper, using MIT Lincoln Laboratory test data KDDCUP99, it is specifically used for intrusion detection evaluation. We use a subset of the data set on 10% as the experimental data source, which contains a total of 494,021 network connections, which normally connected 97277, abnormal connections 396,744.

Here we analysis for data of DOS attack types (DOS attack type coded as "0001").

Let A is sample data matrix where each row vector representation of a sample data, the row vector containing 34 data, assuming that there is a records sample, then A is a matrix of 34 rows; due to enter the network after each sample data will have a corresponding output, coupled with the examples of the type of DOS attack with code "0001" indicates, the desired output matrix B is a matrix of a rows and 4 columns (temporary threshold is not considered, only consider the weight issue). In this paper, BP neural network input nodes is 34, the hidden layer nodes is 15, the output layer nodes for four although the connection weights neural network weights can be represented as a real number string, but during the training of the network, also need to be a real number string into two parts, set the input layer to the hidden layer connection weights matrix W1, then W1 is the matrix of 34 rows and 15 columns; Similarly, the hidden layer to the output layer connection weights weight matrix W2 is 15 rows and 4 columns. So we can get the formula (1).

$$A_{a\times 34}W^{1}_{34\times 15}W^{2}_{15\times 4} = B_{a\times 4} \quad (1)$$

In formula (2), A and B is the coefficient matrix, C is the augmented matrix. After elementary row after conversion formula (3) as shown with constraints.

$$C = [A \quad B] \quad (2)$$

$$C = [A, B] = \begin{bmatrix} A' & B' \\ E & D \end{bmatrix} \quad (3)$$

In formula (3), C matrix and D matrix is 0, after subsequent processing, Corresponding to the output from the original A, B now becomes A ', corresponding to the output B', through this process, we can be a large sample into smaller sample, making computing faster, more streamlined sample data. In order to make the sample used in classification sync detector detects the proposed model, we first classify the sample data consolidation, were constructed DOS, PROBE, U2R, R2L four categories attack sample data sets, so that each of the four detectors detection four categories attacks. In order to reduce the number of suspicious attack, that attack the four types of data sets overlap between the number of records to be small. Increase the accuracy of calculation of the amount over the General Assembly, which will reduce the learning speed; precision is too small, will increase the number of records overlap, resulting in a suspicious number of attacks increases, the impact of training results.

Table 1 decimal reserved a table when recording duplication

DOS	44	0	37	4
PROBE	49	37	0	5
R2L	33	4	5	0
U2R	11	1	2	6
	Normal	DOS	PROBE	R2L

Table 2 decimal reserved four recording duplication

DOS	1	0	0	0
PROBE	1	0	0	0
R2L	0	0	0	0
U2R	0	0	0	0
	Normal	DOS	PROBE	R2L

Table 3 sample data compression case

	Before compres sion	**Decimal point one**	**Decimal point four**
Normal	97277	3918	85811
DOS	391459	228	37436
PROBE	4107	233	1775
R2L	1119	52	921
U2R	59	20	52

After optimization of the training samples of using the above method, using matrix elementary row transformation matrix to maintain a linear relationship between the column vectors this conclusion, we can further streamline the sample data, the sample data can be maintained so streamlined between the attribute fields linear relationship. Training the neural network is actually a given sample in real time to adjust the network connection weights through the process, the results of the sample pretreatment convergence for network training play a key role.

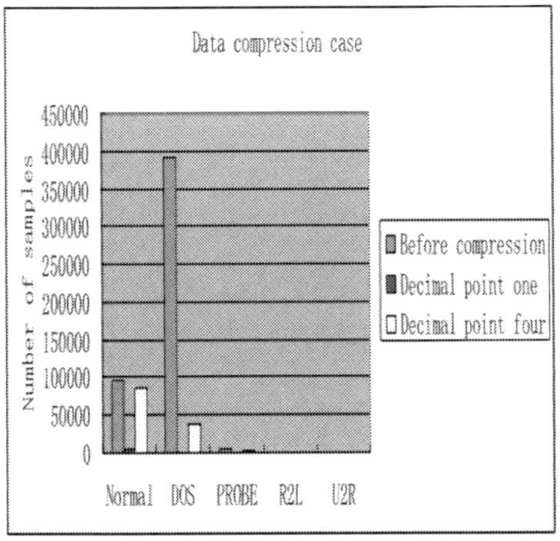

Figure 1 data compression case of FIG.

4 Conclusion

(1) Analysis of neural network nonlinear predictive multivariable predictive superiority and neural network for forecasting disadvantage lies.

(2) The proposed five important parameters affecting network forecasting capabilities. The sample quality, sample normalization, the input layer nodes, hidden nodes and the target value of MSE for network training tolerance.

(3) Training error allowed under certain circumstances, to study the impact on the network without a parameter prediction accuracy, and found there is an optimal sample input layer nodes and hidden nodes, such networks have strong predictive capacity.

(4) This paper is constructed using genetic algorithm neural network while optimizing impact prediction accuracy of the parameters (number of nodes in the input layer, hidden layer nodes and sample allows the training error) algorithm, has been on better network prediction model Finally, numerical examples verify the correctness of the analysis results.

References

[1] Chen Guo analyzed factors predictive accuracy of neural network model [J] Beijing: Pattern Recognition and Artificial Intelligence 2005,18 (5): 528-533

[2] Jiang Lin, Chen Tao, Qu Liangsheng affect the quality of the training sample performance of artificial neural networks [J] Beijing: China Mechanical Engineering 1979,8 (2): 50-53

Binding [3] LI Min-qiang, Xu Bo Yi, Ji-song Kou genetic algorithms and neural network [J] Beijing: System Engineering Theory and Practice 1999 (2):

[4] WU Huai-yu, SONG Yu-chieh neural network nonlinear regression analysis [J] Wuhan: Wuhan University of Science and Technology Metallurgy, 1998,21 (1): 90-93

[5] WANG Yi-huai, Wang Lin. Nonlinear regression based on artificial neural networks [J] Beijing: Computer Engineering and Applications 2004, 12, 79-81

[6] high-Hong Shen, SUPPORTED nonlinear regression artificial neural network prediction model [J] Beijing-based: North China University of Technology, 1999,11 (1), 68-73

Carbothermal reduction process of silica formed from shirasu volcanic ash using solar furnace

Keisuke Hatakeyama [1], Keigo Sato [1] and Kensuke Nishioka [1,a]

[1] *Faculty of Engineering, University of Miyazaki, 1-1 Gakuen Kibanadai Nishi Miyazaki 889-2192, Japan*

Abstract. Metallurgical grade silicon was formed using Shirasu volcanic ash as starting material with solar furnace. The solar furnace was composed of two parts: Fresnel lens and reacting furnace. The reacting furnace was composed of a cylindrical vacuum chamber and quartz glass plate functioning to guide the concentrated sunlight into the furnace, and was placed at the focal point of the Fresnel lens. The sample was made from a mixture of silica formed from Shirasu volcanic ash and carbon, and placed in the carbon crucible inside the reacting furnace. The sample was irradiated for 3 hours, and the furnace was left until it cooled down to room temperature. After the cooling process, the sample was mixed and placed in the carbon crucible, and it was irradiated and cooled with the same processes again. After the experiment, the sample was evaluated by X-ray diffraction and the production of silicon was confirmed.

1 Introduction

A kind of the volcanic ashes called "Shirasu", abundantly deposited due to a big pyroclastic flow in the Southern Kyushu, Japan, which happened 20 to 100 thousand years ago, is one of the unused natural resources. Shirasu is unsuitable for agriculture due to excessive water drainage, and poses a heavy damage every year as sediment disaster. In order to use the Shirasu which may cause sediment disaster and serious damage to the crops, various areas had been investigated to find its promising application [1, 2]. In view of this matter, we provide the formation method of metallurgical grade silicon (MG-Si) from Shirasu volcanic ash, as a starting material.

MG-Si is used as an alloy element in steel industries and metallurgical companies. MG-Si is used for the production of solar cells and electronic devices [3]. All those products are produced on a commercial scale by conventional carbothermal reduction process. The overall reaction of MG-Si production using silica with carbon is shown by the following equation.

$$SiO_2(s) + 2C(s) = Si(l) + 2CO(g) \qquad (1)$$

In industry, MG-Si is mainly produced by silica and carbon at 1900°C [4]. It is an energy intensive process. In recent years, the interest regarding renewable energy has been growing due to the decrease in fossil fuels and the global warming. With a view of growth in the world's population, the demand for energy is always growing, while the fossil fuel reserves, in particular oil, are in the stake of exhaustion [5]. That is why, the use of clean, safe and cost-effective energy supplies need to be increased. Solar energy which has an intensity of 1000 W/m^2, has an important role to play within this framework, as it may also be used in solar furnace application. The solar furnace is possible to produce high temperature in a short period of time by utilizing Fresnel lens as a sunlight concentrator. This method is a new production process to create materials that require to be done under high temperature condition with regards of the shortage of energy resources.

In this study, instead of requisite thermal energy, we prepared a MG-Si using the solar furnace that produced an ultrahigh temperature condition by utilizing a Fresnel lens as a sunlight concentrator. The MG-Si was prepared by silica formed using Shirasu and carbon.

2 Experimental procedure

The solar furnace was composed by two parts; Fresnel lens and reacting furnace. In solar thermal conversion applications, Fresnel lens has the advantages of small volume, light weight, suitable for mass production at low cost, and effectively increasing the energy density. The material of the Fresnel lens was PMMA, and it had an area of 1.40 m × 1.05 m. Figure 1 shows the concentrator system and the schematic of the reacting furnace. The reacting furnace was composed of a cylindrical vacuum chamber (diameter: 20.3 cm, depth: 19.2 cm) and quartz glass plate functioning to guide the concentrated sunlight into the furnace, and we put a thermal insulation and crucible, at the Fresnel lens' focal point.

The sample was made from a mixture of silica formed using Shirasu volcanic ash [6, 7] and carbon, and

[a] nishioka@cc.miyazaki-u.ac.jp

placed in the carbon crucible, as shown in Fig. 2, inside the reacting furnace.

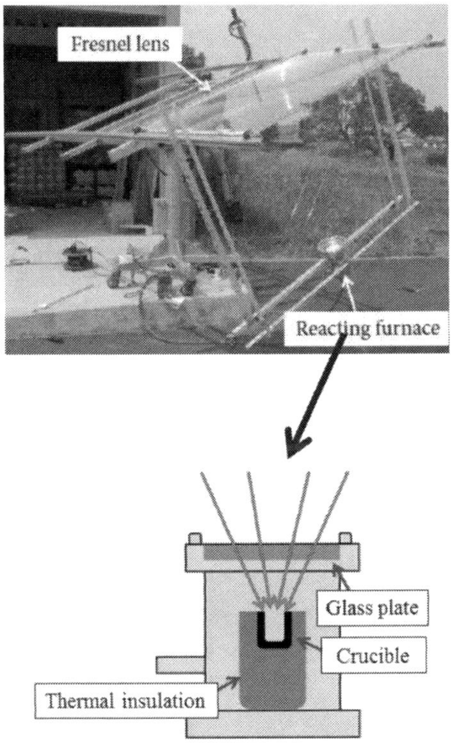

Figure 1. Concentrator system and schematic of the reacting furnace.

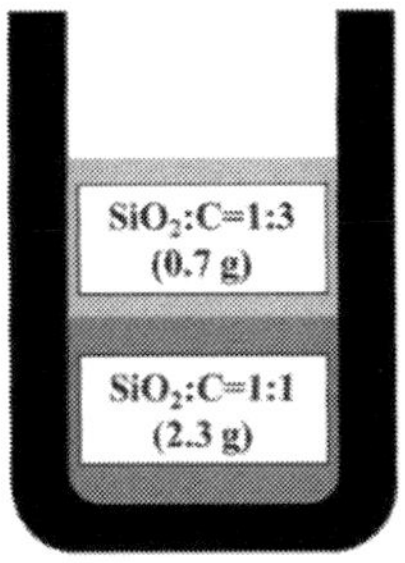

Figure 2. The sample ratio in crucible.

In order to form silica from Shirasu, the materials (Shirasu: 59.6wt%, Li_2CO_3: 4.8wt%, B_2O_3: 23.5wt%, MgO: 5.3wt%, $CaCO_3$: 4.4wt%, Na_2CO_3: 2.4wt%) were mixed using blender for 3 minutes. The resultant powder mixture was placed into a platinum (Pt) crucible and was melted at 1400°C for 3 hours in an electric furnace. The mother glass was formed by quenching the melt in pure water. The bulk particle sizes were adjusted to be less than 1 mm in diameter. Leaching was performed by immersing the mother glass into 1 M hydrogen chloride (HCl) solution at 90°C for 4 hours in total; tha acid solution was refreshed after immersing the mother the glass for 2 hours. The silica was then washed with deionized water and dried.

In the actual process of the MG-Si formation, three reactions take place,

$$SiO_2(s) + C(s) = SiO(g) + CO(g) \tag{2}$$
$$SiO(g) + 2C(s) = SiC(s) + CO(g) \tag{3}$$
$$SiO(g) + SiC(s) = 2Si(s) + CO(g). \tag{4}$$

SiO is easily produced by the reaction (2) near 1100 °C. Then SiC formation occurs through reaction (3) above 1300 °C. In the solar furnace, the upper part of the sample is irradiated with sunlight and becomes the high temperature. Therefore, the sample was prepared as shown in Fig. 2. After that, the air inside the reacting furnace was discharged so that it was in the vacuum condition. Then the Ar gas was left to flow into the reacting furnace until it reaches 0.1 MPa. Next, the sample was irradiated with high concentrated sunlight for 3 hours. While the Ar gas kept on flowing at 2 L/min into reacting furnace. After the irradiation process, the furnace was left until the sample was cooled down to room temperature under Ar atmosphere. The sample was mixed, and placed in the carbon crucible, and was irradiated in the same way again for 3 hours, and was cooled down to room temperature.

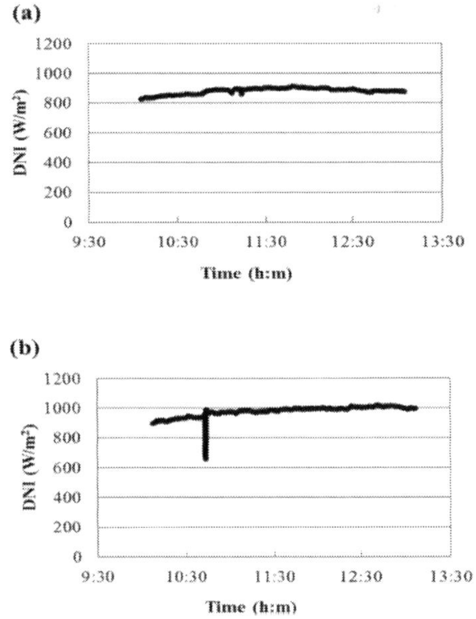

Figure 3. Direct normal irradiance (DNI) during the (a) first and (b) second irradiation processes.

Fig. 3(a) and (b) shows the direct normal irradiance (DNI) during the first and second irradiation processes, respectively. DNI was stable and high with approximately 900 W/m^2 and 1000 W/m^2 during the experiment. The solar furnace system uses the Fresnel lens to collect sunlight and focuses it onto the reacting chamber. The system can only use the direct beam component of sunlight. Since the solar furnace is equipped with the sun tracking system, temperature of the furnace depends on the intensity of DNI. Therefore, it is necessary to irradiate the sample using a stable and high DNI. Finally, the sample was evaluated by X-ray diffraction (XRD, PANalytical, X'Pert PRO).

3 Results and discussion

Fig. 4(a) and (b) shows the image of the sample inside the inner carbon crucible after first and second irradiation process, respectively. From the figure, it was confirmed that the dent ("Irradiated portion") was formed on the sample, where the highly concentrated sunlight hit. At the portion in the Fig. 4(a), green material and glass lump was formed. On the other hand, at the portion in the Fig. 4(b), green and black material was formed. This particular part ("Irradiated portion") of the samples were analyzed using XRD.

Fig. 5(a) and (b) shows the result of XRD pattern of the sample after first and second irradiation processes, respectively. The result of XRD pattern analysis in the Fig. 5(a) showed that the silica, carbon and SiC diffraction peaks were appeared. From the XRD result, the sample was found to have produced no Si. On the other hand, the result of XRD pattern analysis in the Fig. 5(b) showed that the SiC and other four diffraction peaks were appeared. The four diffraction peaks that labeled with (111), (220), (331) and (422) were considered as the characteristic of Si. From the XRD result, the sample was found to have produced the Si.

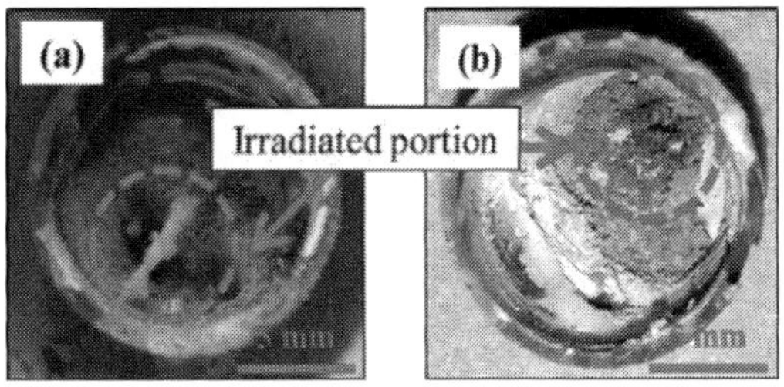

Figure 4. Image of sample inside the crucible after (a) first and (b) second irradiation process.

In the previous study, we have successfully prepared MG-Si from silica and SiC, which were formed from Shirasu volcanic ash, using the solar furnace [8]. DNI at that time was approximately 900 W/m^2. In the case that carbon was used as the reducing agent, the Si wasn't formed at 900 W/m^2. However, the Si was formed after the second irradiation process. This is because SiC was generated and carbon disappeared in the sample by the first irradiation.

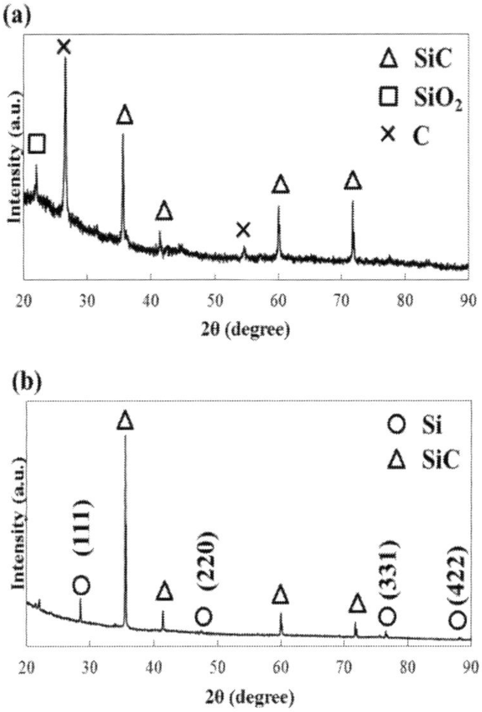

Figure 5. X-ray diffraction pattern of the sample after (a) first and (b) second irradiation processes.

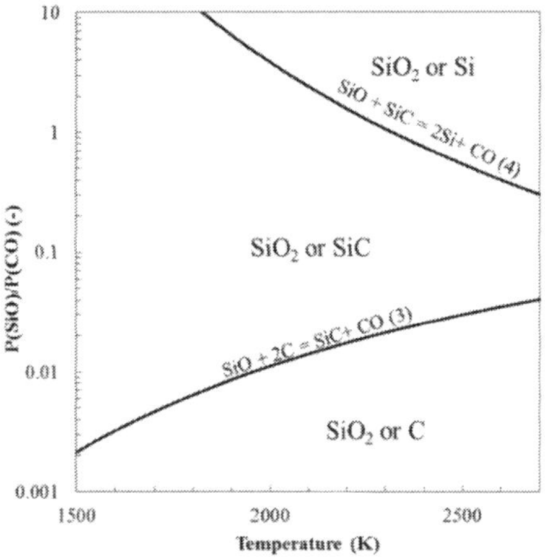

Figure 6. Phase diagrams of the silica reduction process.

DOI: 10.1051/matecconf/20166502001

Fig. 6 shows the phase diagram of the silica reduction processes with carbon calculated using data reported by Nagamori et al. [9]. The phase diagram indicates that reaction (3) is easier than the reaction (4). If SiO_2 and C exist in the reaction system, reaction (3) takes place preferentially.

4 Summary

We have successfully prepared Si from silica, which were fabricated from Shirasu volcanic ash, and carbon using the solar furnace. The solar furnace was composed by two parts; Fresnel lens and reacting furnace. The material of the Fresnel lens was PMMA, and it had an area of 1.40 m × 1.05 m. The reacting furnace was composed of a cylindrical vacuum chamber (diameter: 20.3 cm, depth: 19.2 cm) and quartz glass plate functioning to guide the concentrated sunlight into the furnace, and we put a thermal insulation and crucible, at the Fresnel lens' focal point. The sample was made from a mixture of silica formed using Shirasu volcanic ash and carbon, and placed in the carbon crucible inside the reacting furnace.

The Si was formed after the second irradiation process. In order to form Si after the first irradiation process, higher DNI or SiC as the reducing agent will be necessary. The method implemented in this study could effectively avert the shortage of energy resources in contrast to the traditional process for the formation of Si. This method is expected to be the valid utilization method of Shirasu volcanic ash as a natural resource and to form superior materials using renewable energy.

References

1. T. Yamaguchi, K.Takewaka and S. Mori, Proceedings of the 6th International Conference on Concrete under Severe Conditions **1**, 209 (2010)
2. M. Kukizaki, Separation and Purification Technology **69**, 87 (2009)
3. J.C.S. Pires, A.F.B. Braga, P.R. Mei, Solar Energy Materials and Solar Cells **79**, 347 (2003)
4. B.G. Gribov and K.V. Zinov'ev, Inorganic Materials **39**, 653 (2003)
5. B.G. Gribov and K.V. Zinov'ev, Applied Solar Energy **48**, 84 (2012)
6. K. Sato, T. Kokubu, K. Nishioka, Advanced Materials Research **747**, 547 (2013)
7. K. Sato, T. Kokubu, K. Nishioka, Advanced Materials Research **622**, 970 (2013)
8. K. Hatakeyama, H. Kaneko, K.Nishioka, International Journal of Materials, Mechanics and Manufacturing **4**, 152 (2016)
9. M. Nagamori, I. Malinsky, A. Claveau, Metallurgical Transactions B **17**, 503 (1986)

Geant4 Simulation Study of Dose Distribution and Energy Straggling for Proton and Carbon Ion Beams in Water

Qiang Zhao[1], Zheng Zhang[1] and Yang Li[1]

[1]*Beijing Key Laboratory of Passive Safety Technology for Nuclear Energy, School of Nuclear Science and Engineering, North China Electric Power University, Beijing 102206, China*

Abstract: Dose distribution and energy straggling for proton and carbon ion beams in water are investigated by using a hadrontherapy model based on the Geant4 toolkit. By gridding water phantom in $N\times N\times N$ voxels along X, Y and Z axes, irradiation dose distribution in all the voxels is calculated. Results indicate that carbon ion beams have more advantages than proton beams. Proton beams have bigger width of the Bragg peak and broader lateral dose distribution than carbon ion beams for the same position of Bragg peaks. Carbon ion has a higher local ionization density and produces more secondary electrons than proton, so carbon ion beams can achieve a higher value of relative biological effectiveness.

Keywords: Monte Carlo simulation, dose distribution, geant4, Bragg peak.

1 Introduction

Early in the forties of last century, Wilson suggested that using the beams of charged particles such as protons and light nuclei for treating malignant tumors [1]. Compared with photons, therapy with protons or heavy ions (such as carbon ions) have advantages in that the beneficial dose profile with a sharp dose fall-off at the end of the particle range [2]. This special advantage makes that killing tumors without destroying healthy tissues becomes possible. Furthermore, an additional advantage of carbon ion irradiation is its increased biological effectiveness in killing tumor cells close to the Bragg peak. Until now proton and ion beams of intermediate energies have become widely used for cancer treatment [3-10]. Many proton therapy facilities and several carbon ion therapy facilities are constructed or are being constructed [11, 12]. However, carbon ions may have a small amount nuclear fragmentation with medium [13-15]. Doses from secondary nuclear fragments can not only enhance the biological effect, but also have unexpected energy deposition beyond Bragg peak. This feature makes a very selective impact on the tumor possible, and it requires thorough treatment planning based on more reliable calculation. Moreover, long-duration missions on the Earth's moon or exploration of other plants in solar system will be more frequent in the future. One has to pay more attention on space radiation which comprised of high energy protons and high charge and energy nuclei [16]. Further investigations on radiation effects of heavy ions, especially on the spatial effects, are needed to understand the risk of cancer and other diseases from the space radiation. The radiation of ion beams is so essential that the energy deposition and biological effect distribution of ions should be adequately described. In this work, we use the Geant4 toolkit to calculate dose distribution and energy straggling of proton and carbon ion beams in water (a tissue-like medium).

2 Model and simulation details

Geant4 is a toolkit for the simulation of the passage of particles through matter. Its areas of application include high energy, nuclear and accelerator physics, as well as studies in medical and space science [17]. In our simulations, the choice of models considers both electromagnetic interactions and hadronic interactions. To ensure the accuracy of our calculations of the depth-dose distribution, the maximum step size in track calculations is taken to be as small as 0.01 mm during the entire event. In addition, we set a cut value of 0.01 mm in the whole detector region.

We construct the pure water phantoms with the water density of 1.0 g•cm^{-3}. During simulation, a typical event is as follows. Firstly, the beam goes through a beam line which includes range shifters, collimators, etc. Then the primary kinematics consists of a single proton or $^{12}C^{6+}$ particle which hits in a (40 mm)^3 water cube perpendicularly. The incident energy for proton is 34 MeV, 62 MeV, and for carbon ion is 62 MeV/u, 113.5 MeV/u respectively. To ensure the accuracy of the calculation of the dose distribution, we run 1,000,000 events for every incident energy.

In order to quantify the amount of energy deposited by ion beams in matter, one usually considers the average energy deposition per unit mass expressed in Gy. These characteristics are calculated by gridding a cubic phantom into a matrix of $N\times N\times N$ cubic voxels each with a corresponding lateral dimension and calculating the energy deposited in each of the small cubic voxels. We make a matrix of 200×200×200 cubic voxels each with a lateral dimension of 0.2 mm. By comparison with the maximum step size (0.01 mm), the small cubic voxel is big enough to ensure the accuracy of our simulations. Using above small cubic voxels method based on hadrontherapy model, one can get the dose in every small cubic voxel. Moreover, according to the local effect model (LEM) developed in GSI, the value of Relative Biological Effectiveness (RBE) can be calculated for each voxel of the treated volume.

3 Results and discussion

Figures 1a-1d show the two dimensional (incident direction and transverse direction) dose distributions for proton and carbon ion beam traversing a water phantom at different energies, respectively. We also give dose contours at the bottom for each case. From Figures 1a and 1c, we can see that they almost have the same position of Bragg peaks (at about 10.5 mm) for beams of proton with 34 MeV and carbon ion with 62 MeV/u, so do beams of proton with 62 MeV and carbon ion with 113.5 MeV/u (Bragg peaks at about 29.5 mm) in Figures 1b and 1d. This is one reason why we choose these energies for the two kinds of beam, although the main reason is that the two kinds of beam at these energies have been using for hadrontherapy in many facilities [20].

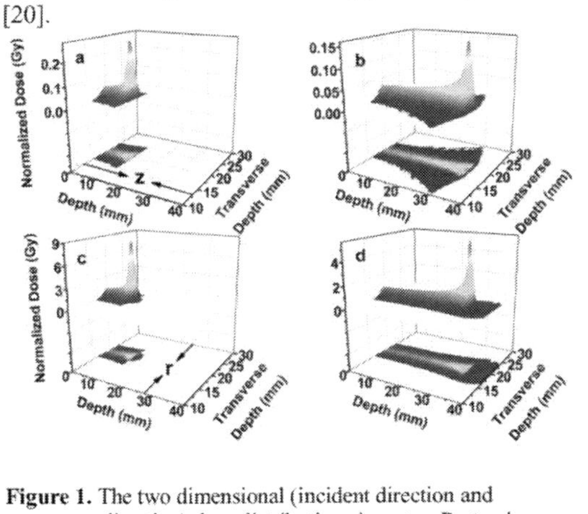

Figure 1. The two dimensional (incident direction and transverse direction) dose distributions in water. Proton beam energy at 34 MeV (a) and 62 MeV (b); Carbon ion beam energy at 62 MeV/u (c) and 113.5 MeV/u (d). The normalized doses (Gy) are calculated in 200×200×200 small cubic voxels, one voxel is 0.2×0.2×0.2 (mm)^3.

The proton beams show more straggling along the incident direction, while the carbon ion beams show distinct tail dose beyond the Bragg peaks. When the carbon ion through water, undergoes a small amount nuclear fragmentation. This process produces a tail of lighter fragments beyond the Bragg peak. The tail reduces the sharp dose contours produced by carbon ion, although for carbon ion it stays within tolerable limits. As shown in Table 1, the dose of Bragg peak is 0.269 Gy for proton beam and 8.56 Gy for carbon ion beam at the same position of Bragg peaks (10.5 mm). The other group is 0.164 Gy for proton beam and 5.58 Gy for carbon ion beam at the same position of Bragg peaks (29.5 mm). The dose of Bragg peak for carbon ion beam is significantly larger than that for proton beam. However, for carbon ion beams, not only the Bragg peaks' width (σ_z) but also lateral width (σ_r) are smaller than that for proton beams when they have the same position of Bragg peak. Figures 2a-2d show the other two dimensional (both transverse directions) dose distributions. We also show dose contours at the bottom for each case. From the figures 2a-2d, except the value of the peaks are very different, the size of dose contours at the bottom for proton beams are slightly bigger than those for carbon ion beams, means that the proton beams have a bigger region of lateral dose distribution as shown in Table 1.

Table 1. Physical quantities of dose distribution for proton and carbon ion beams in water. The position (R), the normalized dose (D), the beam direction width (σ_z), and the lateral width (σ_r) of Bragg peak for H^+ ions and $^{12}C^{6+}$ ions at beam energy (E) are given.

Beam	E(MeV/u)	R(mm)	D(Gy)	σ_z(mm)	σ_r(mm)
H^+	34	10.5	0.269	1.2	3
H^+	62	29.5	0.164	3	3
$^{12}C^{6+}$	62	10.5	8.56	0.8	2.5
$^{12}C^{6+}$	113.5	29.5	5.58	1.8	2.5

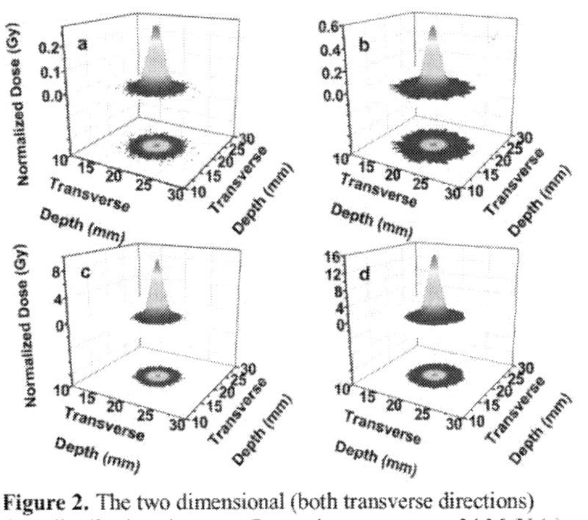

Figure 2. The two dimensional (both transverse directions) dose distributions in water. Proton beam energy at 34 MeV (a) and 62 MeV (b); Carbon ion beam energy at 62 MeV/u (c) and 113.5 MeV/u (d). The normalized doses (Gy) are calculated in 200×200×200 small cubic voxels, one voxel is 0.2×0.2×0.2 (mm)^3.

The lateral scattering draws a lot of attentions because it limits the closest approach of a beam passing a critical structure. To give a clear view of dose distribution in detail, we can give any sectional drawing of each dimension. In Figures 3a-3d, we show a sectional drawing of dose distribution which traverses the position of center of the transverse depth (20 mm) for every case, respectively. The value between maximum and minimum has been divided into 20 stripes between color of red to blue. The outermost blue stripe represents the lowest dose which may be mostly formed by the secondary electrons. As we known, in the case of ions inject into a medium, firstly they will have ionization with the atomic shell of the medium's atoms, so they may get electrons, if it happens, the charge of carbon ion has to decrease and the charge of proton may down to zero. Secondly, for carbon ion, it has nuclear reaction processes with the medium and has a lot of secondary fragments such as neutron, proton, alpha and other charged particles, which are responsible for the energy deposition beyond the Bragg peak, they can continue to interact with the medium and produce the electrons, too. However, after the energy straggling, around the Bragg peak, the carbon ion may still has a higher local ionization density than proton and the energies of the secondary electrons are also large. So the blue regions for carbon ion beams are far bigger than those for proton beams, especially beyond the Bragg peaks. Along the beam direction, for proton beams, the green to red stripes appeared in an extended region. It can be said that the carbon ion beams have more steep Bragg peaks than those for proton beams. In Figures 3e-3h, we give a sectional drawing which traverse the position of centre of the Bragg peak for every case, respectively. Except for the outermost blue stripe, we can also find the diameter of the contour for proton beam is about 6.5 mm, but for carbon ion beam, the diameter is only about 5 mm. This means the carbon ion beams have more steep gradients than the proton beams, especially between the red stripe to green stripe.

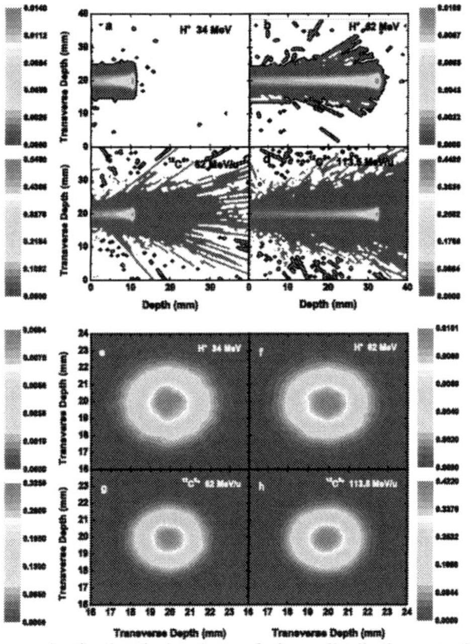

Figure 3. Sectional drawings of dose distributions at the position of center of the transverse depth (20 mm) along the incident direction. Proton beam energy at 34 MeV (a) and 62 MeV (b); Carbon ion beam energy at 62 MeV/u (c) and 113.5 MeV/u (d). Sectional drawings of dose distributions at the position of the Bragg peaks along the transverse direction. Proton beam energy at 34 MeV (e) and 62 MeV (f); Carbon ion beam energy at 62 MeV/u (g) and 113.5 MeV/u (h). The normalized doses (Gy) are calculated in 200×200×200 small cubic voxels, one voxel is 0.2×0.2×0.2 (mm)^3.

4 Conclusion

In summary, proton beams have broader lateral dose distribution and relatively bigger energy straggling along the beam trajectory than carbon ion beams. For carbon ion, it has a higher local ionization density, so can produce more secondary electrons than proton along the incident direction. All these calculated results and analysis may shed light on application of ion beams for cancer therapy. And this method can also be used to calculate dose distribution in many fields in relation with proton and heavy ion radiations.

Acknowledgments

This work was supported by the National Natural Science Foundation of China under Grant Nos. 11275071 and 11305061, and the Fundamental Research Funds for the Central Universities under Grant No. 2014MS53.

References

[1] Wilson R. R., Radiological use of fast protons, Radiology, **47**, pp. 487-491, (1946)
[2] Geissel H. & Scheidenberger C., Slowing down of relativistic heavy ions and new applications, Nucl. Instrum. Meth. B, **114**, pp. 136-138, (1998)
[3] Schimmerling W., Vosburgh K. & Todd P., Interaction of 3.9-Gev nitrogen ions with matter, Science, **174**, pp. 1123-1125, (1971)
[4] Schimmerling W., Miller J., Wong M., et al., The fragmentation of 670A MeV neon-20 as a function of depth in water, Radiat. Res., **120**, pp. 36-71, (1989)
[5] Xu S., Tay B. K., Tan H. S., et al., Properties of carbon ion deposited tetrahedral amorphous carbon films as a function of ion energy, J. Appl. Phys., **79**, pp. 7234-7239, (1996)
[6] Schardt D., Schall I., Geissel H., et al., Charge-changing nuclear reactions of relativistic light-ion beams ($5 \leqslant Z \leqslant 10$) passing through thick absorbers. Nucl. Instrum. Meth. B, **117**, pp. 221-234, (1996)
[7] Amaldi U., Cancer therapy with particle accelerators, Nucl. Phys. A, **654**, pp. C375-C399, (1999)
[8] Kraft G., Tumor therapy with heavy charged particles, Prog. Part. Nucl. Phys., **45**, pp. S473-S544, (2000)
[9] Matsufuji N., Fukumura A., Komori M., et al., Influence of fragment reaction of relativistic heavy charged particles on heavy-ion radiotherapy, Phys. Med. Biol., **48**, pp. 1605-1623, (2003)
[10] Hollmark M., Gudowska I., Belkić, Dž., et al., An analytical model for light ion pencil beam dose

distributions: multiple scattering of primary and secondary ions, Phys. Med. Biol., **53**, pp. 3477-3491, (2008)

[11] Tsujii H., Mizoe J., Kamada T., et al. Overview of clinical experiences on carbon ion radiotherapy at NIRS, Radiother. Oncol., **73**, pp. S41-S49, (2004)

[12] Schulz-Ertner D., Nikoghosyan A., Thilmann C., et al., Results of carbon ion radiotherapy in 152 patients, Int. J. Radiat. Oncol. Biol. Phys., **58**, pp. 631-640, (2004)

[13] Schardt D., Schall I., Geissel H., et al., Nuclear fragmentation of highenergy heavy-ion beams in water, Adv. Space Res., **17**, pp. 87-94, (1996)

[14] Bian B. A., Zhang F. S. & Zhou H. Y., Fragmentation cross sections of 20Ne collisions with different targets at 600 MeV/nucleon, Nucl. Phys. A, **807**, pp. 71-78, (2008)

[15] Zhao Q., Zhang F. S., Wang Z. P., et al., Secondary Beam Fragments Produced by 200 and 400 MeV/u $^{12}C^{6+}$ Ions in Water, Chin. Phys. Lett., **26**, pp. 092501, (2009)

[16] Durante M. & Cucinotta F. A., Heavy ion carcinogenesis and human space exploration, Nature, **8**, pp. 465-472, (2008)

[17] Agostinelli S., Allison J., Amako K., et al., Geant4-a simulation toolkit, Nucl. Instrum. Meth. A, **506**, pp. 250-303, (2003)

[18] Scholz M., Kellerer A. M., Kraft-Weyrather W., et al., Computation of cell survival in heavy ion beams for therapy, Radiat. Environ. Biophys., **36**, pp. 59-66, (1997)

[19] Krämer M., Jäkel O., Haberer T., et al., Treatment planning for heavy-ion radiotherapy: physical beam model and dose optimization, Phys. Med. Biol., **45**, pp. 3299-3317, (2000)

[20] Cuttone G., G. Cirrone A. P., Rosa F. Di., et al., Proton therapy detector studies under the experience gained at the CATANA facility, Nucl. Phys. B, **172**, pp. 79-83, (2007)

MATEC Web of Conferences **65**, 02003 (2016) DOI: 10.1051/matecconf/20166502003
ICNSCM 2016

ELECTRO CONDUCTIVE ALUMINA NANOCOMPOSITES FROM DIFFERENT ALUMINA-CARBIDES MIXTURES

Luis A. Díaz[1], Adolfo Fernández[1], Anna Okunkova[2], Washington Solís[2], Pavel Peretyagin[2], Francisco J. Gotor[3], Ramón Torrecillas[1, 2]

[1]Department of Multifunctional Nanomaterials and Nanocomposites, Nanomaterials and Nanotechnology Research Center (CINN-CSIC-UNIOVI-PA), Avda.de la Vega, 4-6, 33940 – El Entrego (Asturias), Spain
[2] Moscow State University of Technology "STANKIN",1 Vadkovskiy per., Moscow, 127055, Russia Federation
[3]Instituto de Ciencia de Materiales de Sevilla (ICMS-US), Sevilla, Spain

Abstract. This work presents the results of an electro-discharge machined ceramic composites consisting of a base non-conductive ceramic component such Al_2O_3, to which is added sufficient amounts of an electro-conductive ceramic nanoparticles such as TiC, TiNC, NbNC, TaNC, and SiC (whiskers) to achieve an electrical resistance of less than about 100W.cm. With these compositions intricate geometries and features as holes, chamfers, slots, angles, changing radii and complex curves can be electro-discharge machined (EDM) into the ceramic body after Pulsed Electrical Current Sintering (PECS) to achieve maximum density and mechanical properties. In EDM, the electrically conductive workpiece or ceramic blank is eroded by electric discharges or sparks which on a small scale generate localized shock waves and intense heat.
The four compositions studied were in the same proportion for all raw materials: alumina 42 (vol %) + conductive material (TiC, TiNC, NbNC, and TaNC) 22 (vol %) + SiC_w 36 (vol %). Processing was carried out mixing raw materials in the suitable proportions in a polypropylene container with zirconia balls and isopropanol media for 72 hours in order to guarantee the homogeneity of the final compositions. The powders were dried and introduced into a PECS furnace for sintering to 1650°C (100 MPa/2 min). The diameters of the pieces obtained were 20 and about 40 mm and 7 mm thickness.
SiC whiskers reinforced electrically conductive ceramic compositions provide a fully dense material with optimal mechanical properties. The capability of electro-discharge machining obtains good surface quality, chip-free edges, dimensional accuracy and complex shapes. The fracture toughness is improved two to three fold over individual ceramic components. Strength and hardness is also increased.
Some composites were tested as a cutting tool to machine IN-718 nickel-base superalloy industrial laminating cylinders. The composites were formed and electro-discharge machined to a standard size cutting insert.

Keywords: Alumina, Silicon carbide whiskers, PECS, SPS - EDM, TiCN-NbCN-TaCN, Titanium carbide

1 Introduction

Materials for cutting tools inserts fall into several well-known categories. These include high speed steels, cast alloys of cobalt and chromium, sintered carbides and ceramic materials such as alumina.
Ceramic tool materials with oxide matrices, particularly alumina, Al_2O_3, are of increasing interest, asthey exhibit high hot-hardness and very high abrasion resistance, thus making it suitable for high speed machining. Ceramic materials are used in especially difficult applications. They have high hardness, chemical inertness and wear resistance even at elevated temperatures. Ceramic cutting tools made of ceramic+carbide, as alumina-titanium carbide composites, have been successful in machining ferrous and non-ferrous alloys. The utility of the ceramic materials may be limited by its low fracture toughness in application where tools tend to foil by fracture, say, in milling or high speed roughing. Main trends in research of ceramic materials are shifting from high purity single phase like Al_2O_3 to multiphase composite ceramics, and nowadays extensive research is done to identify all the combinations which would satisfy the requirements: high hardness, high toughness, high temperature resistance and inertness toward machining part.

The development of new cutting tool inserts is continuous and remains an issue. In particular, great efforts are being made to improve the fracture toughness by the addition of hard particles of silicon carbide whiskers to the Al_2O_3 matrix and the introduction of Ti(CN), Ta(CN), Nb(CN), etc., cermets as an alternative to WC-Co hard metals [1] to enhance its mechanical properties considerably. Themechanochemical process called mechanically induced self-sustaining reaction (MRS), as proposed by Yen et al., [2], is a novel method to produce high purity transition metal carbonitrides from mixtures of elemental transition metal and carbon in nitrogen atmosphere [3-6].The mechanical milling has shown to be a powerful powder processing technique that allows production of cost-effective ultrafine products and extremely homogeneous composite materials [7].

Electro Discharge Machining (EDM) is a good choice when it comes to machining of very hard materials, provided that they have enough electrical conductivity. The main advantage of EDM is the machining of ceramics is its ability to produce high-complexity shapes, independently of the mechanical properties of the material (hardness, brittleness, flexural strength, etc.). In the case of non-conductive ceramic components it is necessary to add sufficient amounts of an electro-conductive ceramic above the percolation threshold such as metallic carbides, nitrides, etc., to achieve its electrical conductivity and so that it becomes susceptible to EDM [8].

The spark-plasma sintering method (SPS), also known as pulsed electrical current sintering (PECS) or field-assisted sintering technology (FAST),can be roughly compared with the conventional hot press. Additionally a pulsed electric current is applied directly to the graphite mold. The SPS method comprises three main mechanisms of action: a) the application of uniaxial pressure, b) the application of pulse voltage, and c) the resistance heating of graphite dies and sample. It is obvious that the efficiency of this method is influenced by electric and thermal properties of the sample. Consequently, to obtain materials that are dense and nanostructured, it is necessary often to use non-conventional sintering techniques. The initial grain size plays a critical role in the microstructure development during sintering, particularly when the grain sizes are of nanoscale.

In this paper, we present results of the mechanical behaviour of several compositions of a base non-conductive ceramic component such as Al_2O_3 to which is added sufficient amounts of several electroconductive ceramic components such as TiC, TiCN, TaCN, NbCN to achieve an electrical resistance of less than about 100 ohm centimeters specific resistance. To improvethe mechanical properties of these compositions (fracture toughness, thermal shock, etc.) a fixed amount of SiC whiskers was added. The combination of mechanical activation and the SPS has been shown to be suitable for the production of materials having nanostructure and a controlled consolidation level.

2 Experimental procedure

The following raw materials were used as starting powders: 1) TM-DAR grade α-Al_2O_3 powder (Taimei Chemicals Co., Japan) with a specific surface area of 14.6 mgr-1 and an average grain size of 150 nm; 2) Titanium carbide (TiC) from Hubei Minmetals Trading Corporation Limited (China) with an average grain size of 40 nm; 3) XCN (X=Ti, Nb, Ta) also of nanometric grain size synthetized by mechanically induced self-sustaining reaction (MSR) following the procedure of Córdoba, et al. [9]; 4) Silicon carbide whiskers (SiC_w) (diameter = 0.6 µm, length = 10-12 µm) from Tokai (Japan). The morphology of the starting carbonitrides powders is shown in Fig.1.

Figure 1. FESEM of the startingTi, Ta, and Nbcarbonitrides powders.

The processing can be divided into two main steps: 1) with corresponding proportions, alumina (42 vol%), TiC, XCN (X= Ti, Ta, Nb) (22 vol%) and SiCw (36 vol%) powders were regardless mixed in a polypropylene container by using zirconia balls for 72 hours in alcohol media (2-propanol, 99% purity, Panreac, Spain), in order to ensure a good homogeneity of the mixtureand 2) alumina powders and the others non-oxide raw materials were mixed together and rolled again in a polypropylene container with zirconia balls in alcohol media for 72 hours. Then the powders were dried first at 60°C and after at 120°C and sieved below 63 microns.

The obtained powders were sintered by Pulsed Electrical Current Sintering (PECS) (FCT System GMBH, HPD25, Germany), using a graphite die under vacuum (10^{-2} mbar) at temperature above 1650 °C. The temperature was measured by using an optical pyrometer focused on the upper part graphite punch, at about 5 mm from the sample, being 100 MPa and 2 min the applied pressure and dwell time, respectively. The tests were carried out at a heating rate of 100°C min^{-1} to reach the maximum temperature. In this way disks of 20 and 40 mm diameter and about 3 and 7 mm height respectively were obtained.

Density of the composites was measured according to the Archimedes method in deionized water. The microstructure was studied by Field Emission Scanning Electron Microscopy (FESEM) (FEI: Quanta FEG 650). The flexural strength (3-point bending test) and fracture toughness were measured in an Instron 8562 testing machine. These measurements were tested by six beams with 3 mm x 4 mm x 15 mm dimensions for each temperature. K_{IC} measurements were performed using the single-edge-notched-beam technique (SENB) (a/w=0,4 and notch radius ≈ 100 µm) following the guidelines described in the ASTM C1421 standard [10]. Indentation tests with a polished material surfaces were carried out inVickers indenter (Buehler Micronet 5103 equipment) applying a load of 200 g with a dwell of 10 s, performing at least 30 indentations per sample.

Thermal conductivity/diffusivity of the samples was measured using two equipments: Diffusivity measurements were carried out by the laser flash method using LFA 457 Microflash (Netzch, Germany). Experiments were done in nitrogen atmosphere at 25, 300, 600 and 900 °C, attending the ASTM E-1461 and DIN EN 821 norms. The experimental data were analyzed by the Cape-Lehman model and "Pulse

Correction" both implemented in the equipment software. The specific heat (Cp) as a function of temperature was calculated using a DSC from TA Instruments (Discovery), equipped with external cooling RCS90 equipment.

The electrical resistivity of the composites was measured by a four–probe method and using the current intensity of 100 mA.

Electrical discharge machining (EDM) tests on composites were carried out by wire EDM, using a tungsten wire of about 0,2 mm diameter and water as dielectric liquid media.

3 Results and discusions

Figure 2 shows FESEM microphographs of the polished surfaces of the sintered conductive-composites.The grain size of each composition is given by the particle size of the starting carbonitridecomponents. The finest microstructures belonging to composition A + TiC + SiCw, while the other carbonitride microstructures with small to large size are those with Ti, Ta and Nb respectively.

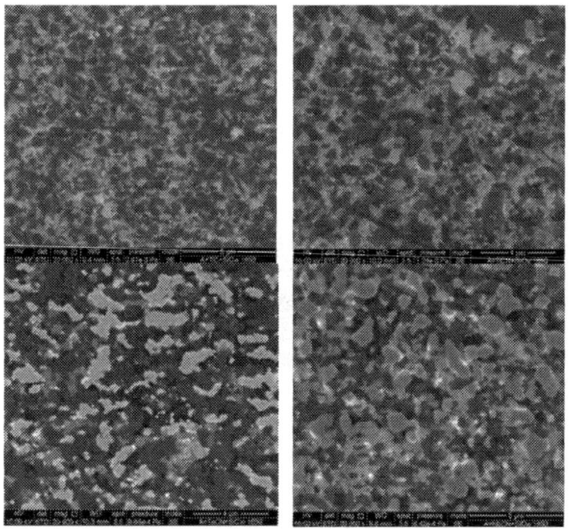

Figure 2. Representative FESEM images of the polished surface of the sintered compositions: A) A+TiC+SiC$_w$, B) A+TiCN+SiC$_w$, C)A+TaCN+SiC$_w$, and D)A+NbCN+SiC$_w$.

The mechanical properties are shown in Table1. Microhardness values for each of the compositions obtained can be observed. The results obtained are correlated to the microstructure obtained and are directly dependent on the particle size.The A+TiC+SiC$_w$composites shown the highest microhardness values whereas lower values occur in the A+NbCN+SiC$_w$composites. As for the flexural strength, higher values are obtained for the compositions of A + TiC + SiC$_w$ (Table 1) and are gradually decreasing with the presence of Ti, Ta and Nbcarbonitridesrespectively. The higher tenacity values are presented in the compositions obtained with alumina, titanium carbide and SiCw. This is linked to the effect produced SiC whiskers distributed on fine matrix of these composites.

Thermal conductivity results are shown in Figure 3.Niobium carbonitrides composites and those with titanium carbide have lower thermal conductivity values at higher temperatures. The values of resistivity and electrical conductivity can be seen in Table 1.

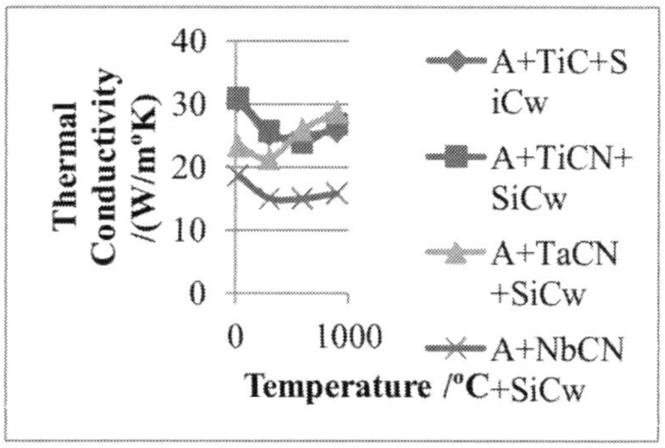

Figure 3. Graphical representation of the thermal conductivities of compositions: A) A+TiC+SiC$_w$, B) A+TiCN+SiC$_w$, C) A+TaCN+SiC$_w$, and D) A+NbCN+SiC$_w$.

Figure 4 shows some of the pieces electro discharge machinedbased on A-TiC-SiC$_w$ and with the Ti and Nbcarbonitrides compositions.

Table 1. Physical properties of the composites elaborates.

Engineering constant		A+TiC+SiC$_w$	A+TiCN+SiC$_w$	A+TaCN+SiC$_w$	A+NbCN+SiC$_w$
Hardness	[GPa]	23,53	22,16	19,34	20,89
Flexural strength (3 points) σ_f	[MPa]	768	451	430	369
Toughness K_{IC}	[MPa.m$^{1/2}$]	7,3	6,57	4,90	6,40
Density	[gr/cm^3]	3,848	3,860	5,427	4,247
Electrical Resistivity	[Ω·m]	9,02E-05	6,66E-05	2,47E-05	1,90E-05
Electrical Conductivity	[S m^{-1}]	1,11E+04	1,50E+04	4,05E+04	1,90E-05

DOI: 10.1051/matecconf/20166502003

Figure 4. A-TiC-SiCwelectro-discharge machining pieces (left).Cutting tools machined by EDMalumina based Ti, and Nb carbonitrides composites (right).

In Figure 5, the microstructures observed by FESEM of the A-TiC-SiC_w specimens are shown. In this Figure, it can also be seen the excellent surface finish of the piece machined by EDM.

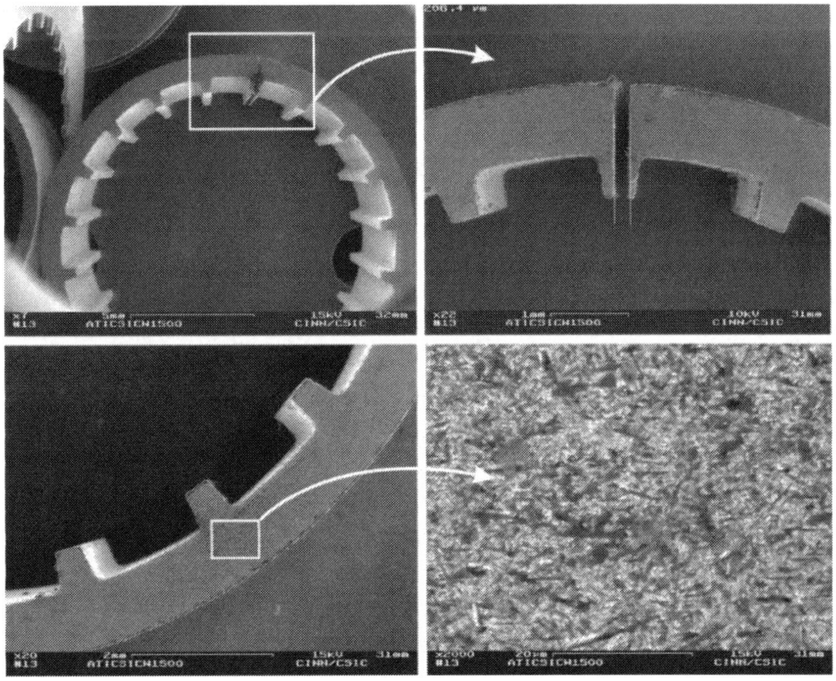

Figure 5. FESEM microphotographies of $ATiCSiC_w$ composition machined by EDM. The thickness of the wire is cutting about 200 microns.

The dimensions of the cutting plates prepared and machined by EDM are shown in Table 2. For engineering evaluation was used to roughing machining TBC cylinders IN-718 nickel-base superalloy, namely type ARAP rolling cylinders (HSS) (Fig.6A), currently being an alternative to high chromium iron (ACRH-N) cylinders because of its better performance, better surface finishes, smaller changes in the mill, and, particularly, reduced machining costs.

The machining of the casting roll was carried out following the same procedure for all the compositions of the prepared plates. Subsequently, rotating the plate and taken other edge, the machining was extended. The aim wastomachineof a length of 250 mm of the roll under these conditions: speed S = 82,5m / min and cutting feed F = 0.55mm / rev. The test performed under these conditions show a different behavior depending on the composition of the plates. Some better than others, and demonstrate good toughness with the continuous presence of continuous metal chips (Fig.6C).

Table 2. Dimensions of cutting plates.

RAW MATERIAL	LENGTH (mm)	WIDTH (mm)	THICKNESS (mm)
Alumina + NbCN + SiC_w	25.40	25.50	5.12
Alumina + TiC + SiC_w	25.34	25.33	7.89
Alumina + TiCN + SiC_w	25.55	25.40	5.96
Alumina + TaCN + SiC_w	26.44	26.60	5.86

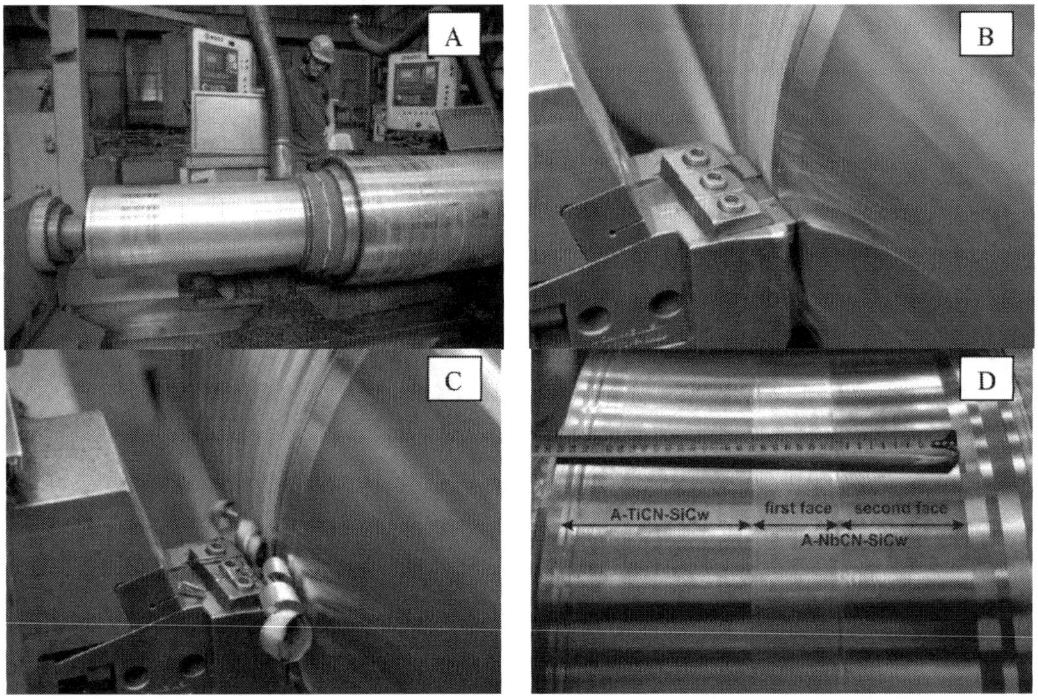

Figure 6. Cilinder rolling tested with the composites elaborated.

In Figure 6, it can be seen different aspects of machining of the cylinder roll tested; it can be observed that the best performances are obtained in the materials developed with titanium and niobium carbonitrides (Fig.6D). In this case, with niobium, it was possible to rotate the specimen and to take advantage, of using another cutting face to continue machining the cylinder avoiding its breakage. This improved behavior could be related to the low thermal conductivity values of both composites. Other compositions showed a quick deteriorationas soon as they suffer a small stroke of the cylinder.

4 Conclusions

Several composites with the same proportions were selected according to the following composition premises: raw materials no conductive (alumina), carbide (Ti) and carbonitride(Ti, Ta, Nb) raw materials and silicon carbide whiskers. The electrical conductivity of the sintered pieces was taken into account in the mixture proportions for composites that could be electro-discharge machined (EDM). The SPS sintering was shown to be an effective method for the fabrication of dense bulk nanostructured materials. The bestmechanical behaviourin the terms offlexural strength and toughness correspond to A-TiC-SiC_w composites and the other compositions have lower values related to their densities.

Alumina-based composites made with TiC and NbCN are the most interesting materials for industrial machining cylinder rolls for steel plants.

This work was supported by the Ministry of Education of the Russian Federation by Contract no. 14.577.21.0089 from 22.07.2014, unique identifier of contract RFMEFI57714X0089.

References

1. L.Chen, W.Lengaver, and K. Dreyer, Advances in modern nitrogen containing hardmetals and cermets, *Int. J. Refract. Met.Hard. Mater,* **18**, 2000, pp. 153-161.
2. B.K. Yen, T. Aizawa, and J. Kihara, Synthesis and formation mechanisms of molybdenum silicides by mechanical alloying, *Meter.Sci. Eng. A*, **46**, 2001, pp. 1-184.

3. J.M Córdoba, M.J. Sayagués, M.D. Alcalá, and F.J. Gotor, Synthesis of titanium carbonitride phases by reactive milling of the elemental mixed powders, *J. Am. Ceram. Soc.*, **88**, (7), 2005, pp. 1760-1764.
4. J.M Córdoba, M.J. Sayagués, M.D. Alcalá, and F.J. Gotor, Monophasic nanostructured powders of Nb, Ta, and Hfcarbonitride phases synthesized by mechanically induced self-propagating reaction, *J. Am. Ceram. Soc.*, **90**, (2), 2007, pp. 381-387.
5. J.M Córdoba, M.D. Alcalá, M.A. Avilés, M.J. Sayagués and F.J. Gotor, New production of TiC_xN_{1-x}-based cermets by one step mechanically induced self-sustaining reaction: Powder synthesis and pressureless sintering, *J. Eur. Ceram. Soc.*, **28**, 2008, pp. 2085-2098.
6. J.M Córdoba, E. Chicardi, and F.J. Gotor, Development of multicomponent-multiphase materials based on $(Ti,Ta,Nb)C_xN_{1-x}$carbonitride solid solutions, *Chemical Engineering Journal*, **192**, 2012, pp. 58-66.
7. C. Suryanarayana, Mechanical alloying and milling, *Prog. Mater. Sci.*, **46**, 2001, pp. 1-184.
8. C. Suryanarayana, Mechanical alloying and milling, *Prog. Mater. Sci.*, **46**, 2001, pp. 1-184.
9. J.M Córdoba, E. Chicardi, R. Poyato, F.J. Gotor, V. Medri, S. Guicciardi, and C. Melandri, Spark Plasma Sintering of TixTa1-x,$C_{0.5}N_{0.5}$-base cermets: Effects of processing conditions on chemistry, microstructure and mechanical properties, *Chemical Engineering Journal*, **230**, 2013, pp. 558-566.
10. ASTM C1421 (2010) Standard Test Methods for Determination of Fracture Toughness of Advanced Ceramics at Ambient Temperature.

Comprehensive analysis of internal and surface defects of ceramics

Sergey N. Grigoriev[1] and Marina A. Volosova[1]

[1]Moscow State University of Technology "STANKIN",1 Vadkovskiy per., Moscow, 127055, Russia Federation

Abstract. System approach to studying of causes of infringement of operational characteristics tool ceramic materials on the basis of complex researches and systematization of their volume and superficial defects is presented in article. The main directions of improvement of tool ceramics are allocated

Keywords: ceramics, cutting tool, fragile destruction, diamond processing, durability, structural defects, coatings.

1 Introduction

The research conducted in the field of studying of the causes of destruction of tool ceramics, show that the main cause of operability loss of ceramics in the process of operation is brittle fracture. The failures caused by brittle fracture account up to 55% of all failures of ceramics when turning and up to 80% when milling.

The reason for this are the low strength properties and responsivity of ceramics to cyclic loads and thermal shocks, which are the result of volume and superficial defects in the structure of ceramics, formed at different stages of its life cycle.

2 Structural features of tool ceramics

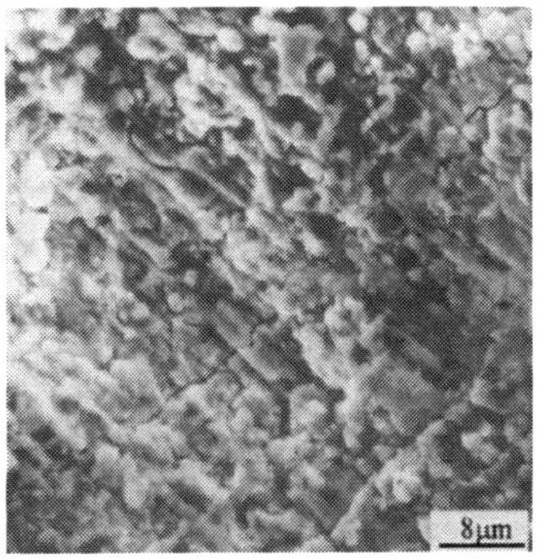

Figure 1: Formation of cracks on the face of tip.

Applying Griffiths failure theory on ceramic tool materials, taking into account the availability in their volume of a granular structure and structural heterogeneity, we can conclude that in the volume and on the surface of the cutting plate there are always conditions of formation and development of cracks. Physical objects for formation of cracks are the interfaces of the grains and inclusions of another chemical composition (Fig. 1). These boundary surfaces are stress concentrators and the source groups of points, from which begins the origin and development of cracks in the volume of the ceramic die (Fig. 2).

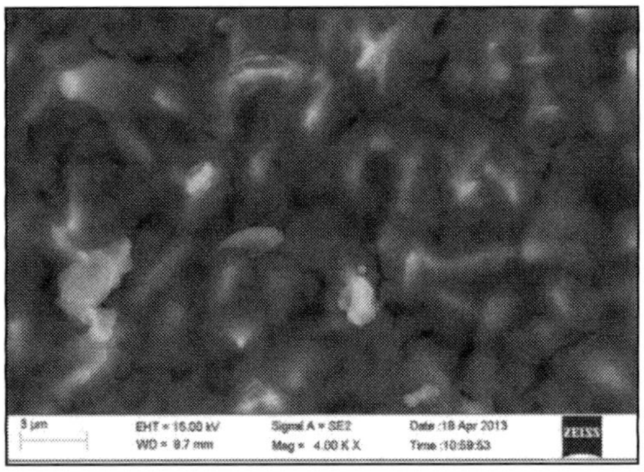

Figure 2: SEM micrograph of structural heterogeneity of tool ceramics.

Taking into account granular structure and structural heterogeneity of ceramics it is reasonable to make the assumption that the weakest points of the ceramic die are the interfaces between ceramic grains and between the grains and inclusions of different chemical composition [1]. At the interface between the grains is formed the front of inner stresses, concentrate the vacancies, arise nonequilibrium thermodynamic potentials. In the unloaded ceramics these forces are balanced in the volume of ceramic die with interatomic interaction forces. The grains in the ceramic die are arranged randomly and contact each other under

 DOI: 10.1051/matecconf/20166502004

developed and arbitrarily located in space surfaces. Therefore, in the same way at the grain boundaries will be aligned the microcracks in the ceramic die.

3 Defects of tool ceramics arising in the process of production

The reduce of porosity is an important issue of technology of manufacturing of ceramics, because the pores are the concentrators of stress and vacancies. (Fig. 3). Properties of ceramics are determined by the overall porosity and its type (closed and opened).

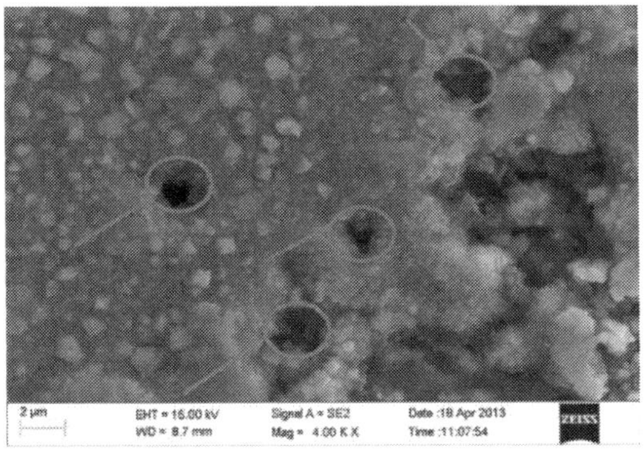

Figure 3: SEM-microphotography of the voids on the surface of the tool ceramic.

By origin it is possible to classify the following types of pores: technological pores formed during pressing: interparticle, intraparticulate, layering pores;

technological pores formed during heat treatment: shrinkage and thermal; secondary, diffusion, coalescence, dislocation, capillary pores.

Intragranular porosity and intergranular porosity are finally formed in the ceramic die as a result of sintering [2]. During the sintering of the ceramic material take place several physico-chemical processes and the sintered product gets certain properties. Thus there is thickening and hardening of the ceramic due to the transfer and redistribution of the material.

The concept of structural heterogeneity for tool ceramics (Fig. 4) is a fundamental physical property and has a wider meaning than for conventional metals and alloys. Under structural heterogeneity of tool ceramics should be understood not only the presence of inclusions of different chemical composition with boundaries in the material volume, but also the presence of structural defects of different types : pores, cracks, cuts, sinks, etc. in the volume of material and on the surface of tool plates.

Figure 4: SEM-microphotography of the structural imperfection of the tool ceramic.

4 Defects of tool ceramics arising in the process of finishing treatment

The most effective method of final machining of tool ceramics are diamond grinding and lapping. The processes that occur during stock removal in diamond grinding, have a significant impact on the operational characteristics of the cutting plate. (Fig. 5). The defining role of defects in the formation of the mechanical properties of ceramics is generally accepted [3].

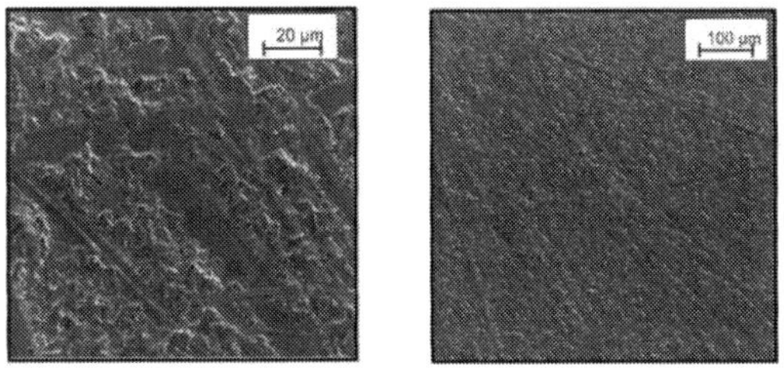

Figure 5: Typical view of the surface of oxide-carbide ceramics after diamond grinding.

In the favorable modes of grinding ceramic plate is not in contact with a post of diamond grinding wheel and the cutting grain is embedded into the material by 0.2-0.4 of the protruding part. The analysis of the process of finishing of the tool ceramics allows to note the following important features.

a) Not all of the grains of the ceramic material will escape from the machined surface of the plate, but only those that protrude above the surface on more than $0{,}5W_l$. Here W_l is the average grain size of the ceramic material.

b) After the tearing out of grain of ceramic material at the place of its location will be formed a crater, having the size and shape of the inverse-depth parts of the grain in Table 1. As the size of the in-depth part of the grain in our model is always less then the size of the protruding one, as the result of grinding the roughness of the machined surface will decrease. The depth of the single crater is always less than half of the nominal size of the ceramic grain.

c) The number of torn grains of ceramic material and the number of craters on the treated surface depend on the homogeneity of the structure and granularity of tool ceramics.

Another characteristic defect of finishing are the so-called grinding microcracks. Practice shows that microcracks form on the surface of ceramics even in the soft modes of grinding, and it reduces the level of its mechanical strength. The destruction of the machined ceramics is usually initiated by a critical surface defect (the weakest link). Another typical defect are marks of the grinding wheel (grinding scratches) (Fig. 6) remaining on the surface of the cutting plate after finishing treatment. The negative effect of grinding scratches on the properties of the ceramics is that traces of the grinding wheel, being in size, in particular, in depth, much bigger than the grains of ceramics, create on the surface of the plate the estimated sources and trajectories for the formation and development of crack front during operation of tool ceramics.

Figure 6: SEM-microphotography of the grinding wheel marks on the face of tip from the mixed CC650 ceramic.

5 Defects of the tool ceramics arising in the process of operation

The study of the tool ceramics from the point of view of structurally inhomogeneous material allows to notice that even in the initial state ceramics contains microcracks, which are localized at the grain boundaries of the material. The grains in the ceramic die are arranged randomly and are in contact with each other under developed and arbitrarily located in space surfaces. Similarly, in an arbitrary order along the grain boundaries appear microcracks in the ceramic die. Under cyclic loading conditions and the intermittent nature of the cut (for example, when milling) on the surface of the cutting plates appear cracks (in the literature these cracks often called planar fractures), resulting in a failure of ceramics in the form of brittle fracture [1]. A sufficiently reasonable assumption that can be made is that these cracks are a development of existing microcracks, which are located at the grain boundaries.

Another typical defect of ceramics, manifested in the course of operation is thermal cracking. The cutting process is accompanied by severe plastic deformation of the processed material and friction at the contact surfaces of the tool, resulting in a large amount of developed heat. Ceramic tools are used typically at high speeds, leading to the appearance of high local temperatures on the front and back surfaces of the tool. Thermophysical properties of ceramic tool (heat capacity, thermal conductivity and thermal expansion) determine the conditions of the thermal load of the cutting edge associated with the balance of heat flows into the cutting tool, chips and processed material, as well as the ability of the cutting tool to resist thermal stresses without failure.

6 Systematization of the defects of tool ceramics

Using the analysis of works of reputable scientists, as well as studies conducted in MSTU "STANKIN", of systematization of surface and volume defects of tool ceramics was based on the technological principle of formation of a defect at a certain stage of the life cycle – in manufacturing (press molding and sintering), finishing (diamond grinding and polishing) and operation in Table 1.

Table 1. Systematization of surface and volume defects of the tool ceramics.

Processes	Type of defect
Production of ceramics	-Extraneous inclusions -Conglomerates of structural components -Structural heterogeneity -Intergranular porosity -Intragranular porosity
Finishing of ceramics	-Torn out grains -Grinding cracks -Burn marks -Marks of the grinding wheel -Residual abrasive
Operation of ceramics	-Impact loading cracks -Thermal cracks -Abrasive wear -Adhesive wear -Chemical corrosion

For example, among the defects arising in the process of pressing and sintering of ceramics are intragranular porosity, intergranular porosity, structural heterogeneity, conglomerates of structural components, inclusions and contaminants. Characteristic defects of finishing of tool ceramics are torn out grains, grinding cracks, marks of the grinding wheel, burns and residual abrasive material in the microroughnesses. Typical defects arising during operation processes of the cutting tool, are impact loadings and thermal cracking, abrasive and adhesive wear and chemical corrosion.

7 Conclusion

The study provides an opportunity to create a system of scientific views on the problems of improving and expanding of the field of application of the ceramic tools. All areas of development and improvement of tool ceramics are associated with the improvement of its mechanical properties, mainly due to minimization of defects formed at different stages of its life cycle.

Most of these defects are formed in the surface layer of ceramics in manufacturing and operation. This fact demonstrates the need for the development and application of technological methods of targeted exposure on the surface layer of the ceramics by coating and/or modification of tool ceramics, which is used after its sintering and shaping, i.e. operations in which appear technological defects.

The work is performed with financial support of the Russian Science Foundation within the Agreement No. 14-29-00297 of August 6, 2014.

The work is carried out on the equipment of the Center of collective use of MSTU "STANKIN" with financial support of the Ministry of Education and Science of Russian Federation, the Agreement No. 14.593.21.0004 of 04.12.2014, the unique identifier of the project RFMEFI59314X0004.

References

1. N.Y. Cherkasova, M.A. Volosova Improving of the Performance of Ceramic Cutting Plates by Mechanical and Ion-Plasma Surface Treatment. Russian Engineering Research. V. 33. 7 (2013), 429-432.
2. V.V. Kuzin, S.N. Grigoriev, M.A. Volosova Effect of a tic coating on the stress-strain state of a plate of a high-density nitride ceramic under nonsteady thermoelastic conditions. Refractories and Industrial Ceramics. V. 54. 5 (2014), 376-380.
3. V.V. Kuzin, M.A Volosova, M.Y. Fedorov Wear of tools from nitride ceramics when machining nickel-based alloys. Journal of Friction and Wear. V. 34. 3. (2013), 199-203.

MATEC Web of Conferences **65**, 02005 (2016) DOI: 10.1051/matecconf/20166502005
ICNSCM 2016

Energy Demand and Environment Integration Model and Application: the Case of Beijing

Guo Fangyuan [1,a]

[1] *State Grid Jilin Power Supply Company, China*

Abstract. In this paper, energy consumption sector is divided into final energy consumption sector and transformation sector, where final energy consumption sector is divided into six sub-sectors, namely, primary industry sector, industry sector, construction sector, transport sector, service and commerce sector and household sector. Particularly, as a case of urban transport, this paper simulates two scenarios: baseline plan and high plan, and respectively sets three sub-scenarios, namely, developing public transport scenario, fuel replacement scenario and improving vehicle fuel efficiency scenario. LEAP model is used to analyze energy demand and environmental emissions in Beijing till 2025.

1 Introduction

Beijing is the capital of China, which has large energy consumption. But it is also a city with scarce energy resources and has high dependence on the external. The study of its energy demand and environmental emissions, in particularly energy demand of urban transport, will provide a scientific basis for planning rational energy development strategy and optimizing energy structure.

There are many complete energy models internationally. Top-down energy model reflects viable technology, but underestimates the potential technological progress [1], such as CGE model [2]. Mixed energy model involve a broad range of technology, and therefore there must be a large number of professional researchers and adequate time to complete the complicated project, such as NEMS model [3]. Bottom-up energy model includes LEAP model [4] and so on.

In this paper, LEAP model is used to study energy demand and environmental emissions in Beijing. The remainder of this paper is organized as follows: the second part describes the sector classification, the third part presents the methodology and scenarios settings; the fourth part is the results of the scenario analysis; and the fifth part outlines the conclusions.

2 Sector classification

The sector classification in China energy statistical system isn't identical to international energy consumption sector classification. In order to make energy scenario analysis better reflect the characteristics of different energy uses and considering the availability of related technical parameters for energy consumption, this paper makes an necessary adjustment in the energy consuming sector, based on China Energy Statistical Yearbook.

The industry sector is the main energy consumption sector. To further study energy consumption in the industry sector and taking into account the rapid development of equipment manufactures, this paper focuses on high energy consumption sectors and equipment manufactures with relatively higher energy consumption in the industry sector. Other sectors with lower energy consumption are merged. After the sector adjustment, the industry sector includes six main sub-sectors: 1) processing of petroleum, coking, processing of nuclear fuel, 2) manufacture of raw chemical materials and chemical products, 3) manufacture of non-metallic mineral products, 4) smelting and pressing of ferrous metals, 5) manufacture of motor vehicles, 6) manufacture of computers, communication equipment and other electronic equipment and others.

Electricity, heat and water production and supply belong to the same statistical category in China statistical system. Electricity and heat production and supply are energy transformation industry, so include the two sectors in the transformation sector. Water production and supply doesn't involve energy transformation process, which is merged with the industry sector.

In China statistical system, the statistics in the transport sector includes only the transportation enterprises engaged in operation. Corresponding energy consumption is only fuel consumption that doesn't include energy consumption of non-operation enterprises and private vehicles. In order to analyze a broad sense of transport energy demand in Beijing, this paper adopts the concept of international general transport. Accordingly, make an adjustment to transport energy consumption data in Beijing Statistical Yearbook. According to specific

[a] guofangyuan0114@163.com

circumstances of each sector, energy consumption in the transport sector includes all gasoline of the primary industry sector, the industry sector, the construction sector, the service and commerce sector and the household sector; 35% of all diesel of the industry sector, the construction sector, the service and commerce sector; and all diesel of the household sector [5]. After the adjustment, the transport sector is divided into the freight sector and the passenger sector. The passenger sector is divided into the intercity passenger sector and the urban passenger sector. The freight sector and the intercity passenger sector are respectively divided into the airway sector, the railway sector and the highway sector. The most important part focuses on the urban passenger sector, which is classified in accordance with public transport (buses and subway), taxies, official cars and private cars.

After the adjustment, energy consumption sector is divided into final energy consumption sector and transformation sector, where final energy consumption sector is divided into six sub-sectors, namely, primary industry sector, industry sector, construction sector, transport sector, service and commerce sector and household sector.

3 Methodologies and scenarios

3.1. LEAP model

LEAP model is energy-environment model based on scenario analysis, which is developed by Stockholm Environment Institute. It can calculate energy demand and environmental emissions.

LEAP model consists of two modules. One is final energy demand analysis module, and the other is energy transformation analysis module. In the final energy demand analysis module, energy demand could be calculated from two factors: activity level and final energy intensity in each sector. The energy transformation module is usually run with the final energy demand analysis module, calculating primary energy consumption to balance secondary energy demand generating in the final energy demand analysis module. The model includes a Technology and Environmental Database that describes the technical characteristics and environmental impacts of a range of energy technologies.

3.2 Regression analysis

Regression analysis is a quantitative analysis method. This method focuses on quantity changes among the variables and describes the relationship in the form of the regression equation.

If the activity level in the transport sector is set to transport turnover, future energy demand is closely related to the changes of transport turnover. The experiences of many countries' economic development show that economic development isn't separable from the development of the transport sector. There is strong positive correlation between transport turnover and GDP or per capita GDP [6]. To forecast future transport turnover, this paper establishes the regression equations between traffic turnover and GDP or per capita GDP by the history data from 1990 to 2012 in Beijing. Econometric software EViews is used in view of its appropriate for analyzing time series.

3.3 Scenarios construction

(1) Population growth

Table 1. Future population in Beijing (million persons)

Population	2015	2020	2025
Total	22.05	24.69	27.05
Urban	19.32	22.22	24.89
Rural	2.73	2.47	2.16

(2) Evolution of industrial structure

Table 2. The percentage of the value-added of every industry accounting for GDP in Beijing (%)

Sector	2015	2020	2025
Primary industry	0.70	0.63	0.52
Industry	22.00	21.05	20.53
Construction	4.16	4.07	3.86
Tertiary industry	73.14	74.76	75.06

Table 3. The percentage of the value-added of every sector (%)

Sector	2015	2020	2025
Petroleum processing	4.05	3.57	3.11
Chemical products manufacture	1.83	1.74	1.60
Non-metallic mineral products manufacture	2.40	2.25	2.03
Ferrous metals smelting	0.20	0.18	0.16
Motor vehicles manufacture	18.50	19.60	20.96
Communication equipment manufacture	9.00	11.50	12.50

(3) The construction of two scenarios

This paper constructs two scenarios. Scenario 1 is the baseline plan. Scenario 2 is the high plan, which mainly forecasts energy demand and environmental emissions with rapid economic development. Meanwhile, three sub-scenarios are respectively setting to analyze energy demand and environmental emissions in different urban transport models, namely, developing public transport scenario, fuel replacement scenario and improving vehicle fuel efficiency scenario.

The basic assumptions of scenario 1: the average annual growth rate of GDP will be 7.5% from 2013 to 2015, 7% from 2016 to 2020, and 6.5% from 2021 to 2025. For the transport sector, in addition to the regression analysis, there are the following assumptions in the urban transport sector: 1) the passenger turnover undertaken by the bus will be 80644 million passenger kilometers in 2025; 2) the passenger turnover undertaken by the subway will account for 50% of the passenger turnover undertaken by the public transport in 2025; 3) continue to limit the number of taxis, maintaining the same number in 2012; 4) the ownership rate of private cars will increase to 0.8 vehicles per household in 2025.

The basic assumptions of scenario 2: the country will put forward a series of policies to promote economic development. The average annual growth rate of GDP will be 7.5% from 2013 to 2025. Other assumptions are the same with scenario 1.

Table 4. Key assumptions of urban transport models

Sub-scenario	Key assumptions of urban transport models
Scenario 1-1 (2-1) developing public transport	By 2025, the average annual transport distance of private cars will drop to 8000 kilometers (km). This part of passenger turnover will be replaced by the subway with the development of the public transport.
Scenario 1-2 (2-2) fuel replacement	By 2025, the percentage of pure electric buses and pure electric taxis will increase to 5%. Develop some pure electric cars. The percentage of pure electric cars will increase to 1%.
Scenario 1-3 (2-3) improving vehicle fuel efficiency	By 2025, the gasoline consumption per 100 km of official cars and private cars will drop to 7 liters per 100 km, with the progressive development of the vehicle manufacturing technology and the knowledge popularity of proper maintenance and use to cars.

4 Calculation results

4.1. Energy demand

Due to space limitation, the data is only about total energy demand in the table. Sectoral energy demand and energy structure in total energy demand will be analyzed in the form of texts.

Table 5. Total energy demand (Mtce)

Scenario	Sub-scenario	2015	2020	2025
Scenario 1	Scenario 1-1	82.66	103.50	127.63
	Scenario 1-2	83.09	104.79	129.92
	Scenario 1-3	82.70	104.05	128.66
Scenario 2	Scenario 2-1	82.66	104.88	133.13
	Scenario 2-2	83.09	106.17	135.42
	Scenario 2-3	82.70	105.11	134.17

According to the calculation results under two scenarios, Beijing's total energy demand will be between 127.63 Mtce to 135.42 Mtce. Compared to the baseline plan, the high plan is with faster GDP growth rate, so energy demand in three sub-scenarios in the high plan will be greater than that in the baseline plan. Energy demand is the minimum in scenario 1-1, 2.29 Mtce less than scenario 1-2, 1.03 Mtce less than scenario 1-3. From 2015 to 2025, the average annual growth rate of energy demand is respectively 4.44%, 4.57% and 4.52% under three sub-scenarios of baseline plan, which is respectively 4.88%, 5.01% and 4.96% in the high plan.

Under developing public transport scenario of the baseline plan, the percentage of energy demand in the industry sector accounting for total energy demand will drop because of the adjustment of industrial structure, which is 22.10% in 2025. The transport sector and the service and commerce sector are the main sectors driving the growth of energy demand. Energy demand in the transport sector will increase to 22.27 Mtce in 2025, the percentage to 22.55%. Energy demand in the service and commerce sector will increase to 48.01 Mtce in 2025, the percentage to 38.61% in 2025. The growth rate of energy demand in the household sector will basically maintain synchronization with that of total energy demand.

From the perspective of energy structure, the demand of natural gas will rapidly increase and coal will decrease, as a result of implementing the policy of natural gas replacing coal. The demand of coke will decrease with the adjustment of industrial structure. Due to the increase of vehicle possession, the demand of gasoline will keep rising. The demand of electricity and heat in the future will still maintain a rapid growth rate with population growth and the development of the tertiary industry. In scenario 1-1, the demand of oil will be accounting for 21.52% of total energy demand, 2.70 Mtce less than scenario 1-2, 1.14 Mtce less than scenario 1-3.

4.2 Environmental emissions

Table 6. Environmental emissions (Mt)

Sub-scenario	2020			2025		
	CO_2	SO_2	NOx	CO_2	SO_2	NOx
Scenario 1-1	109.35	0.12	0.44	127.06	0.11	0.49
Scenario 1-2	111.71	0.12	0.46	131.27	0.11	0.53
Scenario 1-3	109.98	0.12	0.45	128.90	0.11	0.50
Scenario 2-1	110.31	0.12	0.44	130.73	0.12	0.50
Scenario 2-2	112.67	0.12	0.46	134.94	0.12	0.54
Scenario 2-3	110.78	0.12	0.45	132.57	0.12	0.51

As is shown in table 6, environmental emissions of main pollutants except SO_2 will continue to increase, which will bring greater pressure to reduce emissions. However, pollutant emissions per million Yuan of GDP will rapidly decrease because of the constant optimization of industrial structure and energy structure. In the baseline plan, compared with 2012, CO_2 emissions in scenario 1-1 will decrease by 54.47% in 2025, 52.82% in scenario 1-2, and 53.78% in scenario 1-3. In the high plan, those will respectively decrease by 55.80%, 54.24% and 55.14%.

5 Conclusion

Energy demnad and environmental emissions in Beijing are projected to increase under two plans due to economic development and increasing social demand. Scenarios may vary in their growth path in energy demand and environmental emissions, but sectoral energy demand and energy structure in total energy demand won't witness big changes.

The study has shown how the LEAP system, combined with regression models, can provide a valuable tool for assessing future urban transport options in Beijing. The studied scenarios demonstrate that plausible alternative pathways do exist. The method, tool, and examples presented are a starting framework for beginning an opportunity for creative engagement with the future.

References

1. Toshihiko Nakata, Progress in Energy and Combustion Science, **30**, 417 (2004)
2. Y. Gao, Shanghai Environmental Sciences, **27**, 210 (2008)
3. S. W. Hadley, W. Short, Energy Policy, **29**, 1285 (2001)
4. www.energycommunity.org
5. Q.Y.Wang, *Handbook for Energy Conservation*, (Energy Conservation and Environmental Protection Press, Beijing, 2006)
6. Richard Gilbert, Kathleen Nadeau, Transportation and Economic Development Conference, Canada, 2001

Model validation of solar PV plant with hybrid data dynamic simulation based on fast-responding generator method

Dawei Zhao [1,2,a], Dongping Hu[1], Jieqiong He[1], Lei Zhang[1] and Ning Chen [1]

[1] *Renewable Energy Department, China Electric Power Research Institute, Nanjing 210003, China*
[2] *School of Electrical and Electronic Engineering, North China Electric Power University, Beijing 102206, China*

Abstract. In recent years, a significant number of large-scale solar photovoltaic (PV) plants have been put into operation or been under planning around the world. The model accuracy of solar PV plant is the key factor to investigate the mutual influences between solar PV plants and a power grid. However, this problem has not been well solved, especially in how to apply the real measurements to validate the models of the solar PV plants. Taking fast-responding generator method as an example, this paper presents a model validation methodology for solar PV plant via the hybrid data dynamic simulation. First, the implementation scheme of hybrid data dynamic simulation suitable for DIgSILENT PowerFactory software is proposed, and then an analysis model of solar PV plant integration based on IEEE 9 system is established. At last, model validation of solar PV plant is achieved by employing hybrid data dynamic simulation. The results illustrate the effectiveness of the proposed method in solar PV plant model validation.

1 Introduction

With the rapid increase of solar photovoltaic (PV) power integration, its impact on the power grid has received extensive concern and excited wide research interests [1]. The model accuracy of solar PV is the key factor to investigate the mutual influences between it and a power grid. However, this problem has not been well solved. Currently, most research of solar PV model validation is concentrated in solar PV inverter and solar PV unit, and is not enough for solar PV plant or plants.

In recent years, with the wide application of PMU (phasor measurement unit) and WAMS (wide-area measurement system), hybrid data dynamic simulation method is proposed and has been successfully applied in synchronous generator model validation, load model validation, and so on [2-4]. Hybrid data dynamic simulation injects external signals to simulation process and opens traditional dynamic simulation loops for interaction with external signals. It is "hybrid" in the sense of bridging the measurement world with the simulation world [3]. Typical data dynamic simulation techniques consist of fast-responding generator method [2-3], phase shifter method [3], variable impedance method [4] and the ideal voltage source method. Among these methods, the fast-responding generator method injects bus voltage and frequency, and it does not require the phase-angle data (for solar PV plants, rare PMU is installed), so it is more suitable for solar PV plant model validation. However, according to the existing literatures, the implementation details of fast-responding generator method and its application in the model validation of renewable energy power plants are all disclosed, this paper will investigate these problems.

2 Modeling of solar PV plant grid integration system

A solar PV plant grid integration system is designed in this paper, as shown in Figure 1. The system consists of two synchronous generators and one solar PV plant. In Figure 1, the rated active power of synchronous generator G1 is 500MW, and this value for G2 is 700MW. The rate power of solar PV plant is 300MW. The rate active power of Load1 is 300MW, and this value for Load2 is 1100MW.

GE model [5] is used for solar PV plant modeling in this paper, it is realized in DIgSILENT PowerFactory [6] using static generator as interface with the grid, the static generator is a controlled current source, as shown in Figure 2.

[a] Corresponding author: zhaodawei@epri.sgcc.com.cn

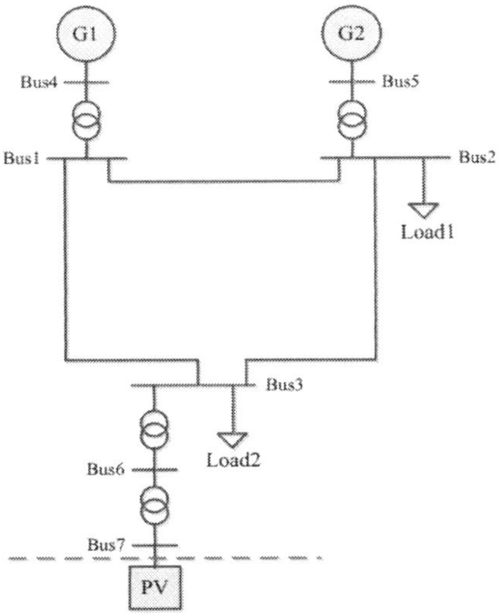

Figure 1. The PV plant grid integration system

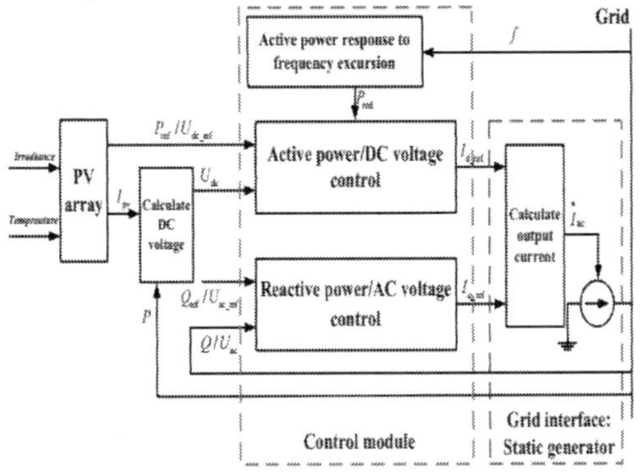

Figure 2. Model structure of PV plant

3 The hybrid data dynamic simulation method based on fast-responding generator and its realization in DIgSILENT PowerFactory

A large synchronous generator with a fast responding exciter and governor is used to inject the external signals. From the terminal of solar PV plant, the large synchronous generator can be seen as an equivalent of the rest of power system, as shown in Figure 3. The generator has very small impedance and transient time constants. A known set of signals (voltage, frequency, real power and reactive power) at Bus7 (see Figure 1 and Figure 3) can be obtained easily from regular measurements. Recorded bus voltage and frequency are input as reference values for the generator exciter (EXC) and governor (GOV) respectively [3]. Because of the large size of the generator and fast response controls, simulated bus voltage and frequency are forced to follow very closely the reference values, i.e. measured voltage V* and frequency f*.

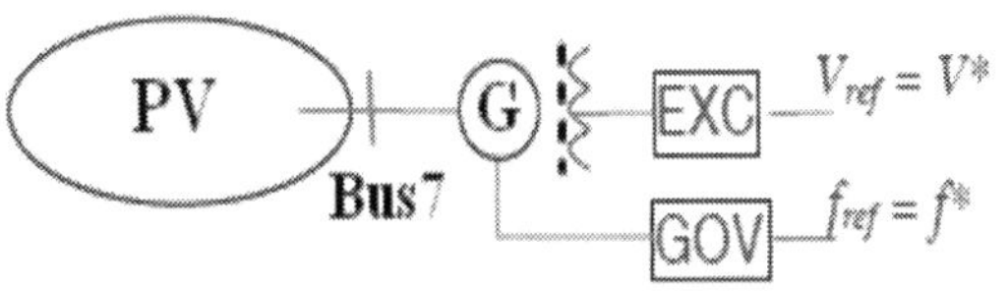

Figure 3. The PV plant grid integration system with a fast-responding generator

The measured values of bus voltage and frequency are all stored in text files, and they can be injected into DIgSILENT PowerFactory software, as shown in Figure 4. In PowerFactory the data can be fed into a generator exciter and governor model using Measurement File (ElmFile) format.

Type1 excitation system (EXC) model and type1 speed governor (GOV) provided by PSASP software [7] are used for synchronous generators in the simulation, the models all have been established in DIgSILENT PowerFactory, and validated by comparing with PSASP [8]. The models of EXC and GOV with injection data are shown in Figure 5.

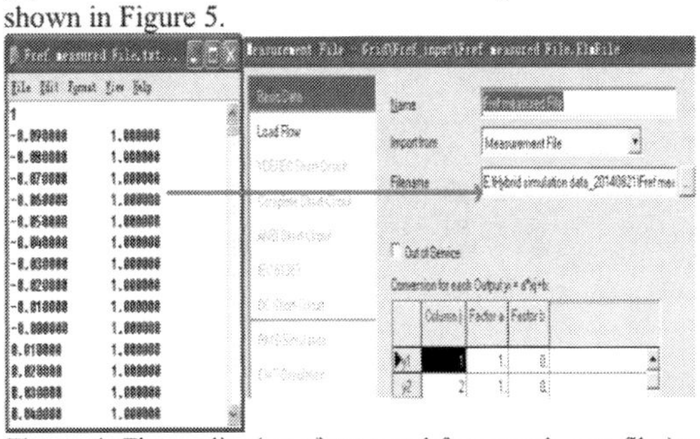
Figure 4. The application of measured frequency's text file in DIgSILENT PowerFactory

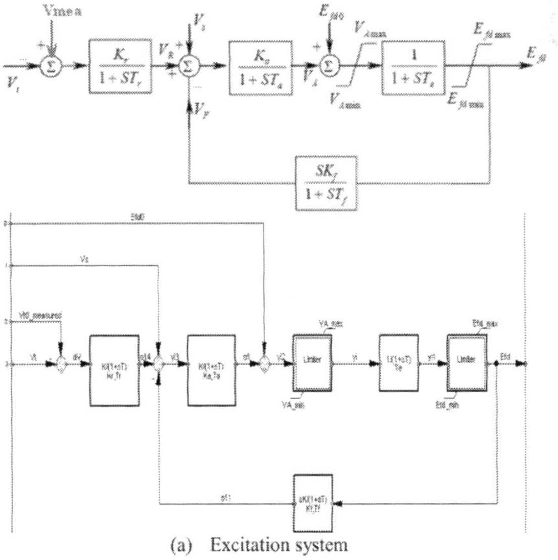

(a) Excitation system

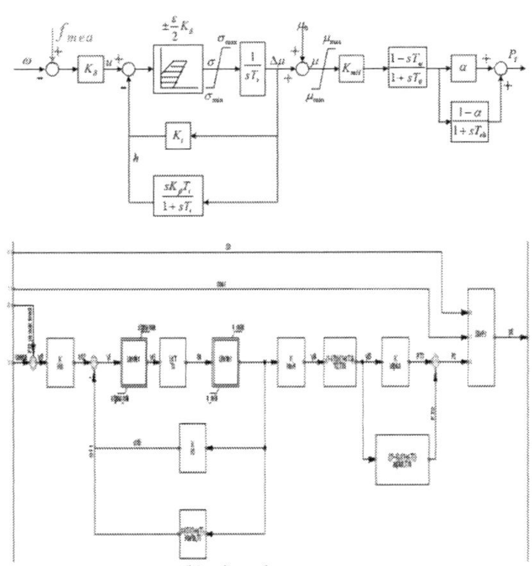

(b) Speed governor

Figure 5. The fast-responding generator's controller realized in DIgSILENT PowerFactory

The parameters of EXC and GOV models used in simulation are listed in Table 1, the time constants are all very small to achieve a rapid control performance. For detailed explanations of the parameters, please refer to [7].

Table 1. The parameters of AVR and PSS models used in simulation.

EXC		GOV		
Kr=1	Kf=0.04	Kd=20	T0=0.001	σ_{max} =1
Tr=0.001	Tf=0.001	K_{mH}=7	Tw=0	σ_{min} =-1
Ka=50	V_{Amax}=10	α =0.333	K_{β} =0	μ_{max} =1
Ta=0.001	V_{Amin}=-10	Ki=0.75	Ti=0.002	μ_{min} =0
Te=0.001	E_{fdmax}=5	Ts=0.001	Trh=0.002	
E_{fdmin}=-5				

4 Simulation results of solar PV plant model validation

Suppose that a three-phase symmetrical short-circuit happens on Bus3 (see Figure 1). This fault begins at 3s and is cleared at 3.1s. According to Figure 3, the measured values of Bus7's voltage and frequency are all stored and injected into PowerFactory as reference values. The comparison between simulation outputs and reference values under this scenario is described in Figure 6.

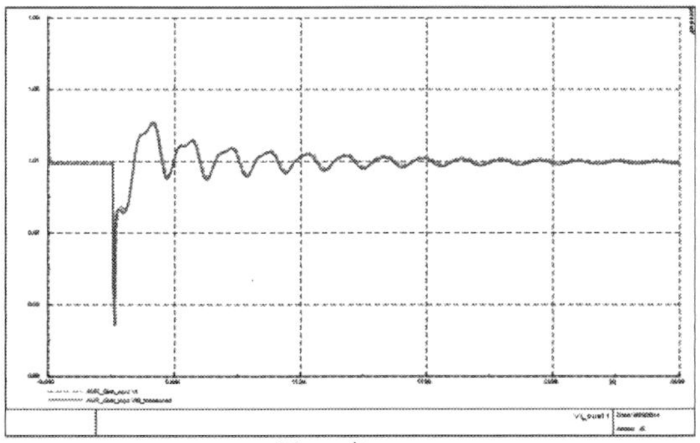

(a) voltage

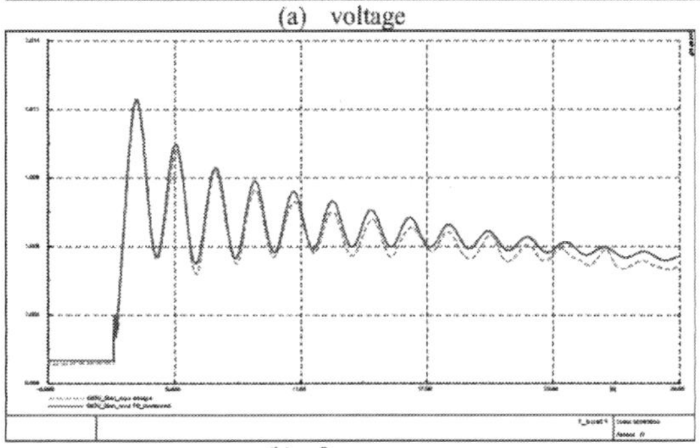

(b) frequency

Figure 6. The comparison between simulation outputs and reference values

The difference between simulation outputs and reference values is acceptable. Then, we can compare the simulated real power and reactive power with the recorded values, if the difference is not small, it demonstrates that PV plant' simulation model or parameter is not accurate enough.

5 Conclusion

This paper presents a model validation methodology for solar PV plant via the hybrid data dynamic simulation based on fast-responding generator method. The implementation details of fast-responding generator method are disclosed for the first time. The simulation results illustrate the effectiveness of the proposed method in solar PV plant model validation.

Acknowledgments

This work is supported by Research Program for youth of China Electric Power Research Institute (Model validation of renewable energy plant using hybrid dynamic simulation method, No. NY84-14-003) and Research Program of State Grid Corporation of China (Renewable power generation performance assessment and fault behavior optimization control technology research and application based on operation data, NY84-14-012).

References

1. S. Eftekharnejad, V. Vittal, G. T. Heydt, et al, Impact of increased penetration of photovoltaic generation on power systems, IEEE Trans. Power Systems, **28**, 2 (2013)893-901
2. D. Kosterev. Hydro turbine-governor model validation in Pacific Northwest. IEEE Trans on Power System s, **19**, 2 (2004) 1144-1149
3. Z. Huang, D. Kosterev, R. Guttromson, et al. Model validation with hybrid dynamic simulation// Proceedings of the 2006 IEEE Power Engineering Society General Meeting: Vol **2**, June 18-22, (2006) Montreal, Canada
4. J. Ma, D. Han, W. Sheng, et al. Wide area measurements-based model validation and its application. IET Gener. , Transm. , Distrib. ,**2**, 6 (2008) 906-916
5. K. Clark, N. W. Miller, and R. Walling, Modeling of GE Solar photovoltaic plants for grid studies, Version 1.1, April 16, (2010)
6. DIgSILENT GmbH, Gomaringen, Germany. DIgSILENT PowerFactory manual version 15.1. (2013)
7. China Electric Power Research Institute, Power System Analysis Software Package (PSASP) user's manual, (2010)
8. D. Zhao, D. Jiang, L. Zhu, et al, Modeling and validation on practical control system of synchronous generator based on PowerFactory, Automation of Electric Power Systems, **38**, 23 (2014) 27-32, 45

The Effect of Adding Carbon Nitride Nanometer Material on The Properties of Epoxy Resin Project

Xing Wang[1,2], Wenliang Sun[1] and Min Guo[2,a]

[1]*School of Economics and Management. Hebei University of Technology, 17 Fukang Road,Nankai District,Tianjin 300191,China*
[2]*Tianjin Academy of Environmental Sciences, 17 Fukang Road,Nankai District,Tianjin 300191,China*

Abstract. Boron nitride(BN) nanometer material was adding as filler to epoxy resin to obtain a high thermal conductive and excellent mechanical properties material. The fillers, which include globular BN, flake BN and fibrous BN, were treated with silane coupling agent and acid to establish a strong affinity with the surrounding matrices. As a result, the thermal conductivity and mechanical properties of the composites were enhanced at a relatively appropriate volume fraction of BN. At a given volume fraction, fillers treated with silane coupling agent provide composites with lower thermal conductivity and higher tensile strength compared with the ones treated with acid. With the increase of the volume fraction of fillers, the thermal conductivity and the tensile strength of the composites show the tendency of increasing, but the tensile yield stress of composites exhibit significant reduction.

1 Introduction

Epoxy resin as the polymer matrix were used most often with reinforcing fibers for advanced composite applications [1]. With the development of epoxy composite material with both high thermal conductivity and excellent mechanical properties in various areas of weight-sensitive aerospace industry, marine, armor, automobile, railway and sporting goods industries, filling inorganic particles of nanometer size with high thermal conductivity and excellent mechanical performance has become a attractive and effective method to strengthen the properties of epoxy resins and meet the higher requirements. Because micro-scale fillers have successfully been synthesized with epoxy resin, BN as a novel nano-scale filler material, which include globular BN, flake BN and fibrous BN, is now being tested to produce high performance composite structures with enhanced properties [1-14]. Recently, boron nitride (BN) based nanofillers such as BN nanoplatelets, BN nanosheets (BNNSs) and BN nanotubes (BNNTs) have attracted increased attention due to its low density, high thermal conductivity, electrical insulation, superb oxidation resistance, passivity to reactions with acids and melts, low coefficient of friction and their potential application in preparing polymer composites for thermal dissipation [15-16].

So far, there have been two approaches developed for nanofiller surface functionalization: covalent and non-covalent methods [2]. The former involves the grafting of reactive modifiers onto the surface of the nanofillers by covalent bonds, while the later is applying the physical interaction between the fillers and modifiers such as van der Walls attractions, electrostatic interactions and hydrogen bonding [17-18].

The purpose of this paper is to show the effect of BN on the thermal and mechanical properties of epoxy [1]. Epoxy resin is chosen as the matrix because of its extremely wide application in aerospace, automotive and electronic industries [2,19]. Zhi et al. prepared a series of polymer nanocomposites with excellent thermal properties by using BNNSs and BNNTs as fillers [20-21]. Previous reports show that various methods have been developed to functionalize BNNSs and BNNTs in order to achieve better processability of their composites . We cured a series of epoxy composites by using different surface functionalization of BN fillers, and the method of the functionalization belongs to covalent functionalization. The BN fillers were firstly treated with silane coupling agent and acid to introduce amine groups and hydroxy groups. The strong covalent bonding formation produced a strong interface between fillers and matrix [2].

2 Experimental

2.1 Materials

The BN powers (globular BN, flake BN and fibrous BN) with an average of 30 μm, concentrated sulfuric acid and nitric acid, ethanol, deionized water, hydrochloric acid, centrifuge, electronic stir bar, disposable straws, beaker, evaporating dish, tweezers, disposable gloves, electronic scale, 50 ml beaker, flake BN, thermometer, electronic mixer, epoxy resin and curing agent.

[a] Corresponding author: 251939587@qq.com

2.2 Treatment with silane coupling agent

Ethanol and deionized water was mixed at the mass radio of 9:1 uniformly (45g of ethanol and 5g of deionized water), 3 drops of hydrochloric acid was added into the mixture to adjust ph value to 4, the silane coupling agent KH–550 at the mass fraction of 1.75% was added into the aqueous solution of hydrochloric acid (7 g), the aqueous solution was placed 3h at room temperature to hydrolysis, then the powers of globular BN were mixed into the solution at the mass radio of 2:7 with KH–550, the mixture was stirred 2h at room temperature and was centrifugal, then it was cleaned with alcohol solution several times until the solution was neutral, finally, the solution obtained was placed in the evaporating dish to dry at the temperature of 50℃ with 3 min and took out, then the solid obtained was ground into a fine powder and collectted to use. The powers of flake BN and fibrous BN were processed in the same way.

2.3 Treatment with acid

The concentrated sulfuric acid and nitric acid was took at volume radio of 3:1 (30 ml of concentrated sulfuric acid and 10 ml of concentrated nitric acid), 1 g of powers of globular BN were added, then the mixture was stirred 3h at the temperature of 50℃ with water bath heating and was centrifugal, then it was cleaned with deionized water several times until the solution was neutral, the solution obtained was placed in the evaporating dish to dry at the temperature of 50℃with 6 min and took out, then the solid obtained was ground into a fine powder and collectted to use. The powers of flake BN and fibrous BN were processed in the same way.

2.4 Adding 1%, 2%, 3% BN powers to EP

20ml of epoxy resin and 0.3g of powers of globular BN was added into the breaker respectively, the mixture was stirred 2h at the temperature of 50℃, then the 10ml of curing agent was added into the breaker, the mixture was stirred 0.5h at the temperature of 50℃ with water bath heating, finally the mixture was encapsulated and cooled at room temperature. According to the methods above, 0.6g and 0.9g of powers of globular BN were added into epoxy resin respectively. The powers of flake BN and fibrous BN were added into epoxy resin at the same volume radio in the same way.

2.5 The preparation of neat sample

20ml of epoxy resin was added into the breaker and the mixture was stirred 2h at the temperature of 50℃, then the 10ml of curing agent was added into the breaker, the mixture was stirred 0.5h at the temperature of 50℃ with water bath heating, finally the mixture was encapsulated and cooled at room temperature.

2.6 Characterization

Tensile tests were performed according to CMT-6104 under a two- end tensile configuration. The tests were conducted in a 10 KN servo-hydraulic testing machine equipped with a Test Ware data acquisition system. The machine was run at a speed of 6mm/min. All the tests were performed at room temperature. Test samples were cut and polished by using NHY-W (universal sample preparation instrument) and XFX (the dumbbell system prototype). Nine specimens from three different materials were prepared for tensile tests.

3 Results and discussion

3.1 The data of tensile properties of composites and data analysis

Typical stress-strain behavior from the tensile tests is shown in Fig.1, Fig.2, Fig3, Fig4. All specimens failed immediately after the tensile stress reached the maximum value. The stress-strain curves showed considerable non-linearity before reaching the maximum stress, and obvious yield point. was found in the curves. Three specimens were tested for each condition.

Table 1. Mechanical properties of neat epoxy and epoxy/BN composites.

	Specimen width (mm)	Specimen thickness (mm)	Modulus (MPa)	elongation at break (%)	Strength (MPa)	tensile fracture stress (MPa)	tensile yield stress (MPa)	yield strain
neat	10.72	3.2	455.75	12.03	31.32	25.23	31.32	12.03
0.5% BN(Acid)	10.76	3.09	459.24	10.13	36.84	35.13	36.84	10.13
0.25% BN(silane)	10.99	3.11	514.8	10.44	39.11	36.46	39.11	9.27
0.25% BN(Acid)	10.36	2.82	462.75	9.77	32.64	28.77	32.64	8.16
0.5% BN(silane)	11.04	3.12	409.14	9.03	33.85	32.97	33.85	9.03
0.75% BN(silane)	10.66	3.08	560.9	8.6	38.47	37.93	17.53	3.02
0.75% BN(Acid)	11.17	2.97	482.22	9.29	36.89	36.88	36.89	9.27

Figure 1. The comparison of the stretchability of neat epoxy and specimens managed (0.25%)

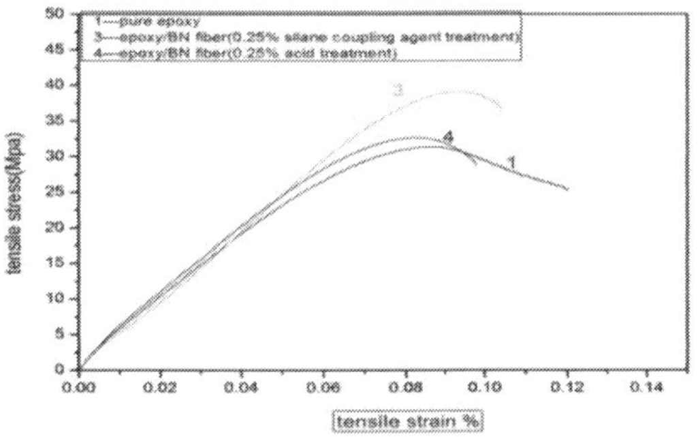

Fig.1showed the stress-strain curve of neat epoxy, composites added BN (acid) at the volume fraction of 0.25% and composites added BN (silane) at the volume fraction of 0.25%. From the curve in Fig.1, by the same tensile stress, we could conclude that the tensile strain of composites added BN (silane) was enhanced slightly, but the tensile strain of composites added BN (acid) do not improved significantly.

Figure 2. The comparison of the stretchability of neat epoxy and specimens managed (0.5%)

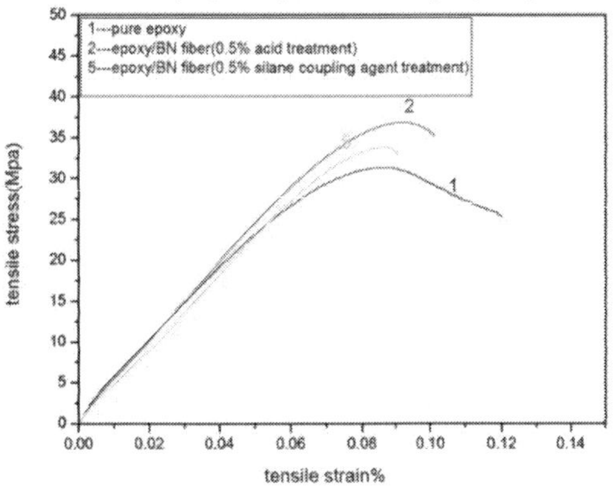

Fig.2showed the stress-strain curve of neat epoxy, composites added BN (acid) at the volume fraction of 0. 5% and composites added BN (silane) at the volume fraction of 0.5%From the curve in Fig.2, by the same tensile stress, we could conclude that the tensile strain of composites added BN (acid) was enhanced slightly, but the tensile strain of composites added BN (silane) do not improved significantly.

Figure 3. The comparison of the stretchability of neat epoxy and specimens managed (0.75%)

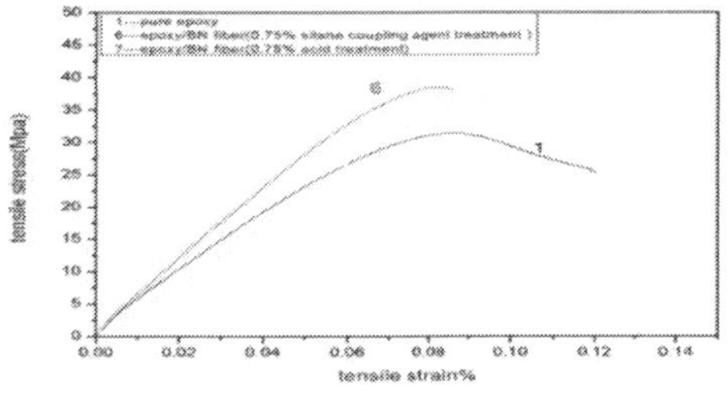

Fig.3showed the stress-strain curve of neat epoxy, composites added BN (acid) at the volume fraction of 0.75% and composites added BN (silane) at the volume fraction of 0.75%From the curve in Fig.2, by the same tensile stress, we could conclude that the tensile strain of composites added BN (silane) was enhanced slightly, but the tensile strain of composites added BN (acid) do not improved significantly.

Figure 4. The comparison of the stretchability of neat epoxy and specimens managed (silane)

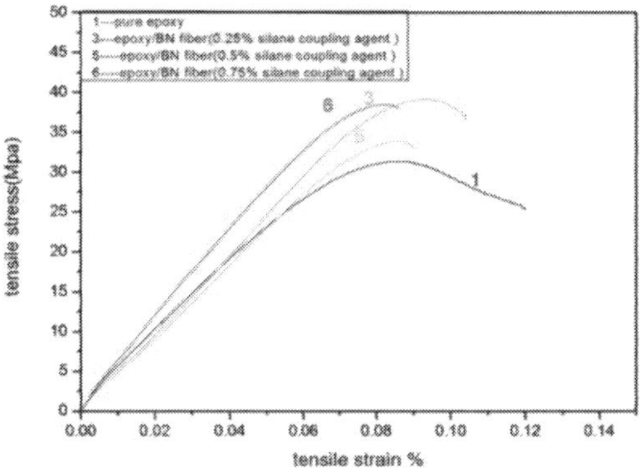

Fig.4showed the stress-strain curve of neat epoxy, composites added BN (silane) at the volume fraction of 0.25% , composites added BN (silane) at the volume fraction of 0.5% and composites added BN (silane) at the volume fraction of 0.75%. From the curve in Fig.4, by the same tensile stress, we could conclude that the tensile strain of composites added BN (silane) at the volume radio of 0.5% was enhanced slightly, but the tensile strain of composites added BN (silane) at the volume fraction of 0.25% and 0.75 do not improved significantly.

Fig.5showed the stress-strain curve of neat epoxy, composites added BN (acid) at the volume fraction of 0.25% , composites added BN (acid) at the volume fraction of 0.5% and composites added BN (acid) at the volume fraction of 0.75%. From the curve in Fig.5, by the same tensile stress, we could conclude that the tensile strain of composites added BN (acid) at the volume radio of 0.5% and 0.75% was enhanced slightly, but the tensile strain of composites added BN (acid) at the volume fraction of 0.25% do not improved significantly.

3.2 Analysis of the tensile performance results

From Table1 and Fig1~3, we could conclude that the tensile strength, elastic modulus, tensile fracture stress and tensile yield stress of the composites materials added silane coupling agent treatment or acidification treatment of BN exhibit enhanced obviously compared with neat epoxy, which proves that BN powers play a role in the composite materials, but the yield strain and the elongation at break decreased, what explain it is that adding BN powers developed the tensile fracture stress of composite materials.

Figure 5. The comparison of the stretchability of neat epoxy and specimens managed (acid)

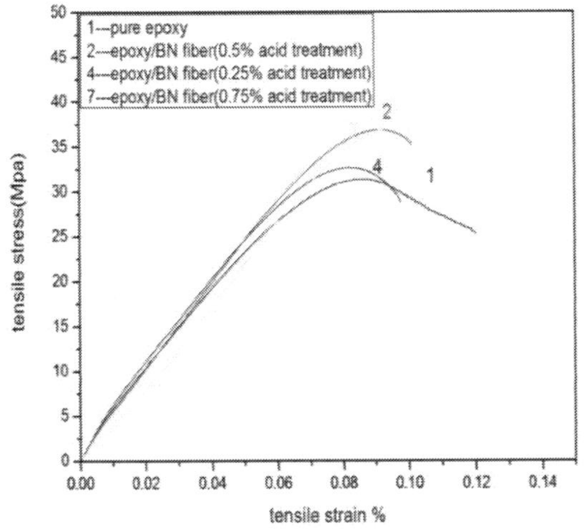

Fig.4 showed that the tensile strength of composite materials added BN (silane) is higher than the tensile strength of composite material added BN (acid) at the same volume fraction, because the structure of BN was destroyed because of acidification, which makes it difficult to perform the net structure in the composites.

Fig.5 showed that the tensile strength of composite material increase continuously and the tensile yield stress of composite material decrease continuously with the growth of BN (acid or silane), the reason is that the net structure changes more and more obviously and the scope of reach of the net becomes bigger and bigger, which makes the tensile fracture stress gets higher and higher.

3.3 Reasons and discussion of thermal conductivity of composite material

From Table3, the thermal conductivity of composites added BN (acid or silane) exhibit enhanced obviously compared with neat epoxy, which proves that BN powers play a role in the composite materials. But the thermal conductivity of composites add BN (acid or silane) at the volume fraction of 0.25% decrease slightly compared with neat epoxy, the reason is that adding BN powers cuts the original chain of thermal conductivity, which not only do not improve the ability of heat conduction but also makes it lower.

The thermal conductivity of composites added BN (silane) is lower than the composites added BN (acid) at the same volume radio of BN powers, the reason is that the acidification enhances the interaction between fillers of BN and matrix, which makes it easy to perform net structure, the net could be used to the faster path to complete the heat dispersion, which develops the thermal conductivity of composites.

The coefficient of thermal conductivity of composite material increase continuously and the tensile yield stress of composite material decrease continuously with the growth of BN (acid or silane), because the area of the net in the composites becomes bigger and bigger, the thermal conductive path is more and more intensive, which is more conducive for the heat condution.

Table 2. The tensile data of the specimens.

serial number	type	parameter (mm)	data		
1	neat	width (W)	10.86	10.83	10.72
		thickness (d)	3.18	3.21	3.2
2	0.5% BN(Acid)	width (W)	10.77	10.76	10.88
		thickness (d)	3.16	3.18	3.09
3	0.25% BN(silane)	width (W)	11.05	10.99	10.97
		thickness (d)	3.11	3.13	3.22
4	0.25% BN(Acid)	width (W)	10.36	10.55	10.4
		thickness (d)	2.87	2.82	2.88
5	0.5% BN(silane)	width (W)	11.06	11.04	11.02
		thickness (d)	3.11	3.12	3.18
6	0.75% BN(silane)	width (W)	10.63	10.69	10.66
		thickness (d)	3.09	3.08	3.2
7	0.75% BN(Acid)	width (W)	11.17	11.3	11.24
		thickness (d)	2.96	3	2.97
Note: color background data for selecting the optimal data during the test					

Figure 6. Coefficient of thermal conductivity contrast before and after processing

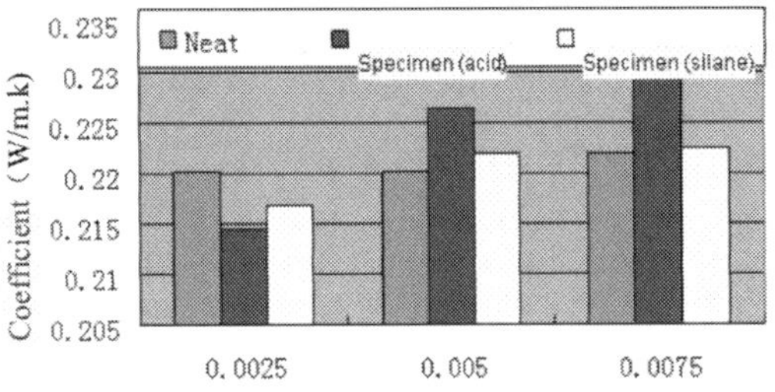

Table 3. The thermal conductivity of epoxy/BN composites.

	environment temperature (K)	**Coefficient of thermal conductivity (W/m*k)**					**optimal data**
neat	296.8	0.2191	0.2206	0.2206	0.2209	0.2211	0.2202
	296.8	0.2192	0.2193	0.2206			
0.25% BN (Acid)	296.56	0.2113	0.208	0.2082			0.2145
	296.74	0.2042	0.2039	0.2044	0.204	0.2041	
	296.93	0.2141	0.2149	0.2146			
	296.29	0.2047	0.2049	0.2047			
0.5% BN (Acid)	297.3	0.2255	0.226	0.2274			0.2263
	297.37	0.2256	0.2271	0.2265			
0.75% BN (Acid)	297.32	0.2298	0.2293	0.2294	0.2298	0.2294	0.2294
	297.55	0.2235	0.2234	0.2239	0.2237	0.2237	
	297.4	0.2136	0.214	0.2143	0.2136	0.2153	
0.25% BN (silane)	296.65	0.2112	0.2115	0.2116			0.2168
	296.83	0.2191	0.2157	0.2156			
	296.86	0.2145	0.2182	0.2172	0.2185	0.2191	
0.5% BN (silane)	295.57	0.216	0.2161	0.2188	0.2155	0.2183	0.2219
	296.26	0.2209	0.2211	0.2218	0.2216	0.2209	
	296.76	0.2203	0.2206	0.219	0.2192	0.2187	
	297.07	0.2243	0.2213	0.2214	0.2221	0.2238	
0.75% BN (silane)	297.67	0.222	0.2223	0.2221			0.2225
	297.81	0.2244	0.2257	0.2252			
	297.82	0.2244	0.2249	0.2246			

4 Conclusions

The tensile strength, elastic modulus, tensile fracture stress and tensile yield stress of the composites materials added silane coupling agent treatment or acidification treatment of BN exhibit enhanced obviously compared with neat epoxy, which proves that BN powers play a role in the composite materials, but the yield strain and the elongation at break decreased, and the thermal conductivity of composites added BN(acid) and BN (silane) at the same volume radio of 0.25 also exhibit the trend of decline, what explain it is that the volume fraction of BN is too little, BN added to the composites could not only not form a network structure to provide channels for thermal diffusion but also destroyed the original heat conduction path.

At the same volume fraction, fillers treated with a silane coupling agent provide composites with lower thermal conductivity and higher tensile strength compared with the ones treated with acid.

With the increase of the volume fraction of fillers, the thermal conductivity and the tensile strength of the composites show the tendency of increase, but the tensile yield stress of composites exhibit significant reduction.

References

1. Yuanxin Zhou, Farhana Pervin, Lance Lewis, Shaik Jeelani, Experimental study on the thermal and mechanical properties of multi-walled carbon nanotube-reinforced epoxy. Material Science and Engineering A 452-453,657-664 (2007)
2. Jinhong Yu, Xingyi Huang*, Chao Wu, Xinfeng Wu, Genlin Wang, Pingkai Jiang*, Interfacial modification of boron nitride nanoplatelets for epoxy composites with improved thermal properties. Polymer **53**,471-480 (2012)
3. Kickelbick G. Prog Polym Sci **28**(1):83-114 (2003).
4. Rozenberg BA, Tenne R. Prog Polym Sci **33**(1):40-112 (2008).
5. Pyun J, Kowalewski T, Matyjaszewski K. Macromol Rapid Commun **24**(18):1043-59 (2003)
6. Chan CM, Wu J, Li JX, Cheung YK. Polymer **43**(10):2981-92 (2002)
7. Maclachlan MJ, Manners L, Ozin GA. Adv Mater **12**(9):675-81 (2000)
8. R. Bagheri, R.A. Pearson, Polymer **41** 269-276 (2001)
9. T. Kawaguchi, R. A. Pearson, Polymer **44** 4239-4247 (2003)
10. H. Mahfuz, A. Adnan, V. K. Rangari, S. Jeelani, B. Z. Jang, Compos. Part A : Appl. Sci. Manuf. **35** 519-527 (2004)
11. M. F. Evora, A. Shukla, Mater. Sci. Eng. A **361** 358-366 (2003)
12. R. Rodgers, H. Mahfuz, V. Rangari, N. Chisholm, S. Jeelani, Macromol. Mater. Eng. **290** (5) 423-429 (2005)
13. P. Farhana, Y.X.Zhou, V. Rangari, S. Jeelani, Mater. Sci. Eng. A **405** (1-2) 246-253 (2005)
14. Y. H. Liao, M.T. Olivier, Z. Y. Liang, C. Zhang, B. Wang, Mater. Sci. Eng. A **385** 175-181 (2004)
15. Golberg D, Bando Y, Huang Y, Terao T, Mitome M, Tang C, et al. ACS Nano **4**(6):2979-93 (2010)
16. Golberg D, Bando Y, Tang CC, Zhi CY. Adv Mater **19** (18):2413-32 (2007)
17. Barus S, Zanetti M, Lazzari M, Costa L. Polymer **50**(12):2595-600 (2009)
18. Sanchez C, Julian B, Belleville P, Popall M. J Mater Chem **15**(35-36):3559-92 (2005)
19. Herman Teo HK, Toh CL, Lu X. Polymer **52**(9):1975-82 (2011)
20. Terao T, Zhi C, Bando Y, Mitome M, Tang C, Golberg D. J Phys Chem C **114**(10):4340-4 (2010)
21. Zhi C, Bando Y, Tang C, Kuwahara H, Golberg D. Adv Mater **21**(28):2889-93 (2009)

Hydrothermal synthesis of $Li_{4-x}Na_xTi_5O_{12}$ and $Li_{4-x}Na_xTi_5O_{12}$/graphene composites as anode materials for lithium-ion batteries

Jiping ZHU [1], Yangyang ZHANG[1], Jie ZHU[1], Rui DUAN[1] and Yumin DA [1]

[1] School of Materials Science and Engineering, Hefei University of Technology, Hefei, Anhui230009, China

Abstract: A potential Lithium-ion battery anode material $Li_{4-x}Na_xTi_5O_{12}$ ($0 \le x \le 0.15$) has been synthesized via a facile hydrothermal method with short processing time and low temperature. The XRD and FE-SEM results indicate that samples with Na-doped are well-crystallized and have more homogeneous particle distributions with smaller overall particle size in the range of 300-600nm. Electrochemical tests reveal that Na-doped samples exhibit impressive specific capacity and cycle stability compared to pristine $Li_4Ti_5O_{12}$ at high rate. The $Li_{3.9}Na_{0.1}Ti_5O_{12}$ electrode deliver an initial specific discharge capacity of 169mAh/g at 0.5C and maintained at 150.4mAh/g even after 40 cycles with the reversible retention of 88.99%. Finally, a simple solvothermal reduction method was used to fabricate $Li_{3.9}Na_{0.1}Ti_5O_{12}$/graphene ($Li_{3.9}Na_{0.1}Ti_5O_{12}$/G) composite. Galvanostatic charge-discharge tests demonstrate that this sample has remarkable capacities of 197.4mAh/g and 175.5mAh/g at 0.2C and 0.5C rate, respectively. This indicates that the $Li_{3.9}Na_{0.1}Ti_5O_{12}$/G composite is a promising anode material for using in lithium-ion batteries.

Key words: $Li_4Ti_5O_{12}$; Na-doping; graphene; hydrothermal synthesis.

1. Introduction

Lithium-ion batteries are a new generation battery that have been rapidly developed over the past few decades. Compared to other second generation batteries, lithium-ion batteries have numerous merits, such as high energy densities, long cycle lifetime, high lithium insertion potential, safety and environmentally friendly, which allows them to be widely used in a range of devices from portable electronics to electric vehicles[1-2]. In contrast with the current commercial graphite anode materials, lithium titanate is a new generation anode material with excellent electrochemical and dynamic properties. Spinel $Li_4Ti_5O_{12}$ has outstanding cycle performance and favorable stable structure due to the zero strain during Li-ion insertion/extraction and not distort in dimensional or volume at high charge/discharge rates. It offers a long voltage-plateau around 1.55V versus Li/Li^+ compatible with polymer electrolytes and high-voltage cathodes, therefore providing an electrode system with much improved safety characteristics and excellent reversibility. However, there are limits to its commercial application due to its intrinsically poor electrical conductivity, sluggish lithium-ion diffusion coefficient and unsatisfactory specific capacity [3-5].

In order to overcome these serious blemishes of $Li_4Ti_5O_{12}$ material, researchers have done a lot of work about these issues. Summarizing the previous reports: three approaches have been used to improve electrochemical performance: (1) explore the effective synthesis route to get nano-sized particles. Nanoscale particles can shorten the Li-ion diffusion distance and significantly broaden contact between electrode and electrolytes [6-11]. (2) introduce the

Foundation item: Projects (No. 21373074) supported by the national natural science foundations of China and (No. 11010202133) supported by Anhui province science and technology plan
Corresponding author: Ji-ping ZHU; Tel:18919665197; E-mail: jpzh@ustc.edu.cn

favorable conductive material into the $Li_4Ti_5O_{12}$, such as a carbon materials or metal powder [10-14]. Recently, Chen et al successfully synthesized a mesoporous $Li_4Ti_5O_{12}$ material that had grown on the graphite oxide (rGO=18wt%). This material possesses impressive high-rate performance with an initial discharge capacity of 132mAh/g at 40C [15]. (3) modified $Li_4Ti_5O_{12}$ with metal or non-metal ions at Li, Ti or O sites to enhance its electronic conductivity , such as Nb^{5+}, Ni^{2+}, Cu^{2+}, Ta^{5+} ,Br^- .

Graphene (G), a two-dimensional macromolecular sheet of carbon atoms with a honeycomb structure, is well known as a novel material in lithium-ion batteries. Two-dimensional graphene sheets with excellent electronic conductivity and mechanical properties have gained interest as a prominent conductive additive for hybrid nano-structured electrodes [16]. Introducing graphene into the $Li_4Ti_5O_{12}$ material via mixing, coating or by growing the $Li_4Ti_5O_{12}$ particles directly on graphene had obtained great progress [14-15]. Zhu et al. [17] synthesized a composite material by implanting graphene into $Li_4Ti_5O_{12}$ nanofibers. Compared to the original $Li_4Ti_5O_{12}$ electrode, this graphene composite presents superior electronic conductivity, long–term cycling stability and outstanding rate performance.

This paper proposes to employ a facial hydrothermal method to synthesize the $Li_4Ti_5O_{12}$ anode material using LiOH and Ti（OC_4H_9）$_4$ as starting materials. Compared to the conventional high temperature solid-state method, this facile hydrothermal route can gain more tiny particles with well crystallinity, and more particles evenly distributed. However, pure $Li_4Ti_5O_{12}$ does not meet the requirements for a competent anode electrode. Herein, different mole ratios of Na have been substituted for Li, and the effect of Na-doping on the electrochemical performance has been investigated. The end result was in order to select the optimum content of Na-doping to improve high-rate performance and electronic conductivity. To our knowledge, modified spinel $Li_4Ti_5O_{12}$ with low content of sodium using hydrothermal method has not been reported until now. Finally, a simple solvothermal reduction method was used to fabricate the optimal anode material $Li_{3.9}Na_{0.1}Ti_5O_{12}$/graphene（$Li_{3.9}Na_{0.1}Ti_5O_{12}$/G）use ethanol as the reductant. This approach can remove the oxygen-containing functional groups (uniformly distribute in the surface of graphite oxide) gradually compared to a chemical reduction method, and has the additional benefit of not producing any toxic byproducts. The composite of $Li_{3.9}Na_{0.1}Ti_5O_{12}$/G fabricated by the solvothermal reduction method has smaller overall particle size and larger specific surface area, and can form a conductive network by means of using two-dimensional structure of graphene, which greatly increase the electronic conductivity and diffusivity of the Li-ion. Electrochemical tests show it has outstanding electrochemical performances, which reveal that this material is a promising anode material for lithium-ion batteries.

2. Experimental

2.1 Synthesis of $Li_4Ti_5O_{12}$ with Na-doping

All raw materials involved in this work were received without any further processing. Pure $Li_4Ti_5O_{12}$ was synthesized via a hydrothermal method with $LiOH.H_2O$ and $Ti(OC_4H_9)_4$ as raw materials. Initially, 17mL tetrabutyl titanate ($Ti(OC_4H_9)_4$) was dissolved into ethanol under vigorously stirred to form a clear solution. A certain quantity of lithium hydroxide monohydrate was dissolved in deionized water to form a 2mol/L LiOH solution. Followed, the LiOH solution was added into the base solution drop and drop with constant stirring. Then , the ivory solution was transferred into a Teflon-lined autoclave and held at 180℃ for 24h. Until the solution cooled to room temperature naturally, washed the precursor with deionized water and ethanol for several times via centrifuging, and lastly dried in an oven for approximately 12h. Finally, the as-prepared precursor was calcinated in a muffle at 700℃ for 8 h with a temperate ramp rate of 4℃min^{-1} to obtain $Li_4Ti_5O_{12}$ powder. Excess $LiOH.H_2O$ (8mol%) was added to compensate the Li source loss in the high temperature calcination process.

In order to obtain the Na-doped samples, LiOH and NaOH solution were added into the base solution simultaneously while undergoing vigorous stirring using a magnetic stirrer, other steps were performed as above. $Li_{4-x}Na_xTi_5O_{12}$ (x=0，0.05，0.1，0.15) samples were gained by adjusting the doping content of NaOH.

2.2 Synthesis of $Li_{3.9}Na_{0.1}Ti_5O_{12}$/G composite

Graphite oxide nanosheets were synthesized by a modified Hummers method using the natural graphite powders as raw material. A 2.0g/L graphene oxide solution

(GO) was prepared by dispersing graphite oxide powder into an ethanol solvent with ultrasonication for 0.5 h. In order to prepare the $Li_{3.9}Na_{0.1}Ti_5O_{12}$/G composite, 1g $Li_{3.9}Na_{0.1}Ti_5O_{12}$ was dissolved in 20mL ethanol and vigorously stirred for 30min, then the previously prepared GO solution was transferred into the $Li_{3.9}Na_{0.1}Ti_5O_{12}$ solution slowly with 3wt% content of GO. The mixture was next transferred into Teflon-lined autoclave and maintained at 120 ℃ for 12h. A gray $Li_{3.9}Na_{0.1}Ti_5O_{12}$/G composite powder was obtained by washing and drying the resulting product.

2.3 Material characterization and electrochemical measurements

The crystalline structure of the prepared samples was identified by X-ray powder diffraction measurements with Cu Ka radiation. A Field emission scanning electron microscope (FE-SEM,SU8020) and Energy Dispersive Spectrometer (EDS) were used to observe the particle morphology and analyze composition of the samples.

In order to test the electrochemical performance of the as-prepared anode materials, anode material, acetylene black, and polyvinylidene fluoride(PVDF) binder at a weight ratio of 8:1:1 was mixed and dissolved in appropriate content of N-methylpyrrolidone (NMP) to form homogeneous slurry. The prepared slurry was then brushed onto a thin aluminum foil substrate using a doctor-blade and dried in a vacuum oven at 120℃ for 12h, then cut into slices with a diameter of 14mm to obtain a working electrode. Pure lithium foil was used as the counter electrode. The cells were assembled in an argon-filled glove box and were tested using CR2032 coin-type half cells. Celgard2400 microporous polyethylene membrane was used as the separator, and 1M $LiPF_6$ dissolved in dimethyl carbonate (DEC)/ethylene carbonate (EC) (1:1 by volume) as the electrolyte. The charge/discharge test was carried out on a CT3008W cell test instrument with a potential range of 1.0-2.5 V under different current densities at room temperature.

3. Results and discussion

3.1 The effect of Na-doping on crystal structure, morphologgy and electrochemical performance

The precursor of $Li_4Ti_5O_{12}$ was prepared through a hydrothermal reaction and calcined at low-temperature to obtain the Na-doped $Li_4Ti_5O_{12}$ powders. The XRD patterns of the as-prepared $Li_4Ti_5O_{12}$ sample with and without Na-doping are depicted in Fig.1(a). The diffraction peaks of all investigated samples can be attributed to expected reflections of spinel $Li_4Ti_5O_{12}$ with a space group of Fd-3m [5]. No impurity peaks were observed, which reveal that a small amount of Na^+ ions have entered the lattice of $Li_4Ti_5O_{12}$ without causing large changes in its structural characteristics. In order to clearly understand the effects of Na-doping in crystal structure, the enlarged peak of (111) plane in the pristine $Li_4Ti_5O_{12}$ and $Li_{3.9}Na_{0.1}Ti_5O_{12}$ samples are shown in Fig.1 (b).

It is easily observed that the $Li_{3.9}Na_{0.1}Ti_5O_{12}$ samples undergo a slight shift toward lower degrees, which demonstrates that the lattice parameter increases due to the Na-doping. This is likely attribute to the larger size of the Na^+ ion（0.102nm）replace the site of Li^+ ion（0.076nm. In addition, the lattice parameter increase is beneficial to enhance the specific capacities of materials.

FE-SEM images of pristine $Li_4Ti_5O_{12}$ and $Li_{3.9}Na_{0.1}Ti_5O_{12}$ are shown in Fig.2. Although all samples are aggregated due to the final calcination process, the $Li_{3.9}Na_{0.1}Ti_5O_{12}$ sample has more homogeneous particle products and more narrow particle-size distributions compared to pristine $Li_4Ti_5O_{12}$. The sample is well crystallized with a particle size in the rang of 300-600nm. In order to gain more information, EDS was carried out to confirm that Na was present in the Na-doped materials. The three different colored bright spots correspond to the presence of Ti, O and Na in Fig.3, respectively. Combined with the above phase analysis, Na^+ ions can be considered conclusively to have entered the lattice structure of $Li_4Ti_5O_{12}$. It can also be noted that all elements are uniformly distributed within the sample.

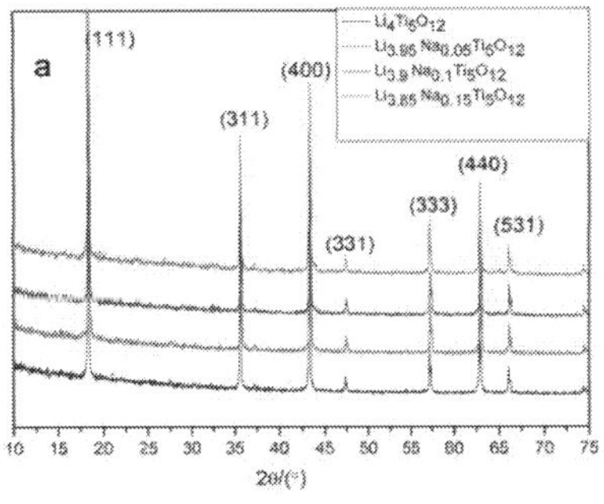

DOI: 10.1051/matecconf/20166502008

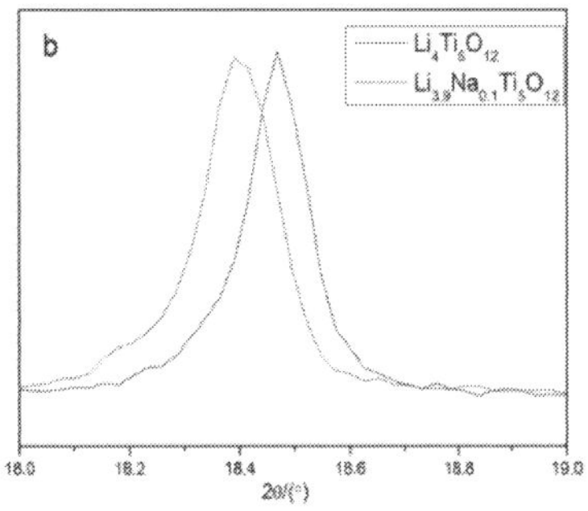

Fig1 （a）XRD patterns of the $Li_{4-x}Na_xTi_5O_{12}$ (x=0，0.05，0.1，0.15) materials; (b) Enlarged (111) peaks of the Li4-xNaxTi5O12 (x=0，0.1) materials

Fig.2 FE-SEM images of $Li_{4-x}Na_xTi_5O_{12}$ (x=0，0.1): （a）x=0，（b）x=0.1

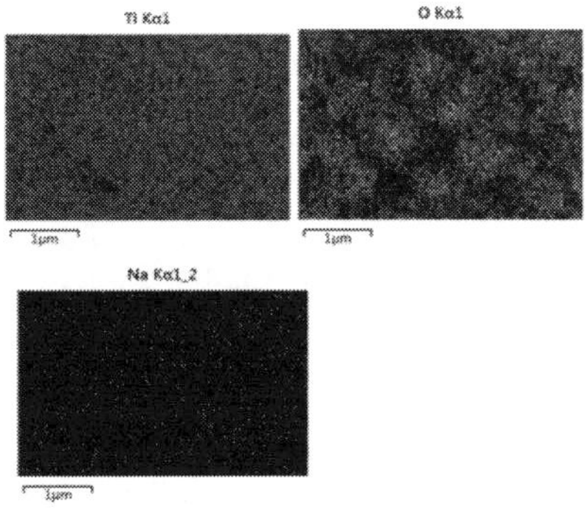

Fig.3 EDS element mapping of $Li_{3.9}Na_{0.1}Ti_5O_{12}$

a

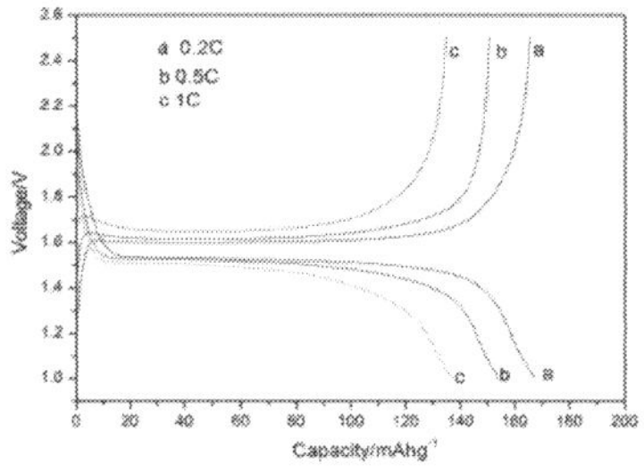

b

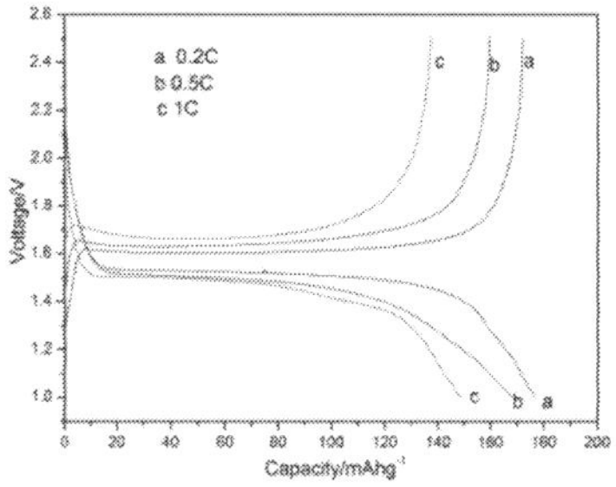

c

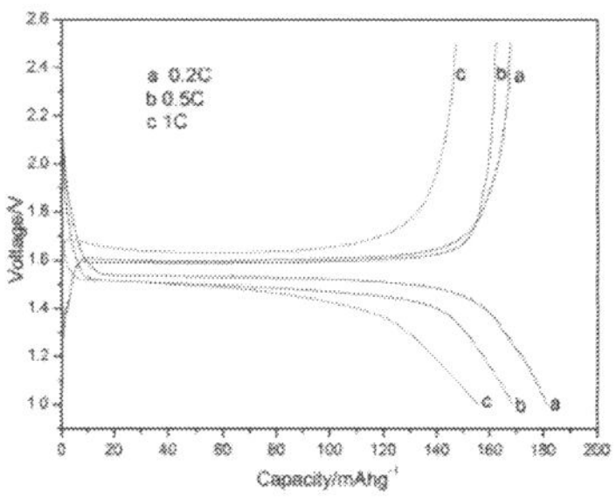

 DOI: 10.1051/matecconf/20166502008

d

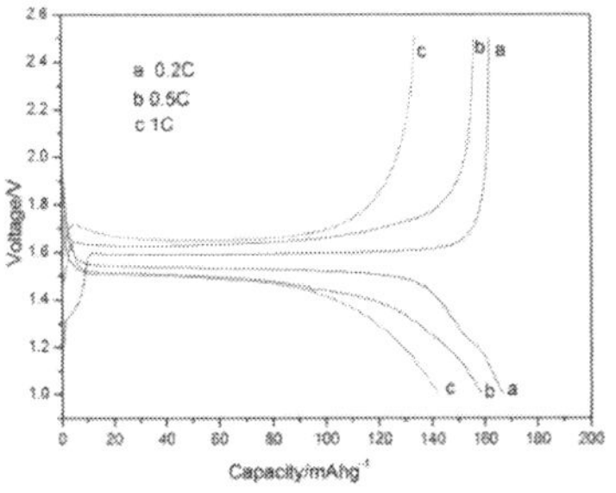

Fig.4 Initial charge/discharge curves of $Li_{4-x}Na_xTi_5O_{12}$ (x=0, 0.05, 0.1, 0.15) samples at 0.2C, 0.5C, 1C rates: (a) x=0, (b) x=0.05, (c) x=0.1, (d) x=0.15

The specific capacities of all samples were tested through the galvanostatic charge/discharge method by a CT3008W cell test instrument. The initial charge/discharge curves of $Li_{4-x}Na_xTi_5O_{12}$ (x=0, 0.05, 0.1, 0.15) samples at different current rates are shown in Fig.4. The $Li_4Ti_5O_{12}$ electrode exhibits a two-phase transition between $Li_4Ti_5O_{12}$ and $Li_7Ti_5O_{12}$ during Li-ion insertion/extraction reactions. All anode electrodes with and without Na-doping have a stable voltage platform at current rates of 0.2C or 0.5C, and the voltage platform occupies 80% of the discharge curve with $Li_4Ti_5O_{12}$. This reveals that a stable $Li_4Ti_5O_{12}$ material can be successfully synthesized by means of a facile hydrothermal method. In addition, the results consistently confirm that Na-doping does not alter the structure of spinel $Li_4Ti_5O_{12}$.

The initial specific discharge capacity of $Li_{4-x}Na_xTi_5O_{12}$ (x=0, 0.05, 0.1, 0.15) electrodes at 0.2C rate were measured as 167.4, 176.7, 181.9 and 166.9mAh/g, respectively. All samples have a high capacity and close to or higher than theoretic capacity. This may be due to the Li-ions have enough time to move between anode and cathode electrodes and the reaction motivation can keep pace with the charge current at low rate, resulting in a lower electrode polarization and higher capacity

The specific capacities of the samples decrease when the charge/discharge current density increases, the effect is particularly noticeable in the pristine sample. Fig.4 shows that the initial specific discharge capacities of $Li_{4-x}Na_xTi_5O_{12}$ (x=0, 0.05, 0.1, 0.15) electrodes at 0.5C rate are 154.0, 168.3, 169.0 and 158.6mAh/g ,respectively, with the $Li_{3.9}Na_{0.1}Ti_5O_{12}$ electrode has the highest capacity among all samples. When the discharge rate increases to 1C, the capacity of the pristine electrode decreases to 137.1mAh/g, but the $Li_{3.9}Na_{0.1}Ti_5O_{12}$ sample retains at 155.7mAh/g, a significant improvement over the pristine sample. This reveals that Na-doping plays an important role in the improvement of reversible capacities at high rates. This may be due to the comparatively large size of the Na^+ ions (0.102nm) replace the Li^+ ions (0.076nm) and lead to the lattice parameter increase, which in turn may lead to an increased in electronic/ionic conductivity and reversible specific capacity. However, the large ion radius of Na^+ ions substitute Li^+ ions may hinder the ion insertion. When the concentration of Na^+ is too low or too high, hindering effects may play a relatively leading role. Therefore, it is important to achieve the appropriate ratio to enhance the electrochemical properties of $Li_4Ti_5O_{12}$.

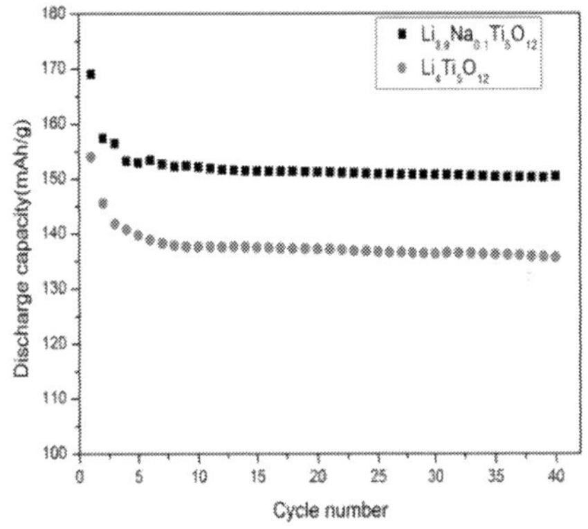

Fig.5 Cycling performance of the $Li_4Ti_5O_{12}$ and $Li_{3.9}Na_{0.1}Ti_5O_{12}$ samples at 0.5C-rate

Cycling performance of $Li_4Ti_5O_{12}$ and $Li_{3.9}Na_{0.1}Ti_5O_{12}$ samples at a current rate of 0.5C are shown in Fig.5. The initial discharge capacities of $Li_4Ti_5O_{12}$ and $Li_{3.9}Na_{0.1}Ti_5O_{12}$ have sharply decreased in comparison to the second discharge process, but their capacities become increasingly steady after cycling several times. This phenomenon may be due to the fact that the first cycle is an activation process or a certain degrees of electrolyte decomposing during the charge/discharge process. The result reveals that the initial discharge capacity of $Li_{3.9}Na_{0.1}Ti_5O_{12}$ electrode maintains at 169.0mAh/g at 0.5C, and the reversible capacity is retained at 150.4mAh/g even after 40 cycles with a capacity retention of 88.99%, higher than the pristine samples of 135.6mAh/g. Based on the above results， Na-doping has been found to result in outstanding cycling stability and excellent high-rate capability.

3.2 Crystal structure, morphology and electrochemical performance of $Li_{3.9}Na_{0.1}Ti_5O_{12}$/G sample

The XRD patterns of the pristine sample, $Li_{3.9}Na_{0.1}Ti_5O_{12}$ and $Li_{3.9}Na_{0.1}Ti_5O_{12}$/G samples are depicted in Fig.6. The XRD patterns of $Li_{3.9}Na_{0.1}Ti_5O_{12}$/G with 3wt%GO are completely consistent with $Li_4Ti_5O_{12}$. As no graphene peaks were found, this indicates that the addition of graphene has not altered the crystal structure of $Li_4Ti_5O_{12}$, perhaps the GO content is below the limit of detection by XRD.

Fig.7 shows the SEM images of $Li_{3.9}Na_{0.1}Ti_5O_{12}$/G composites at different magnifications. Graphene sheets are uniformly distributed among the $Li_{3.9}Na_{0.1}Ti_5O_{12}$ particles and efficiently hindered particle regrowth, which lead to the composite have homogeneous powder and narrow particle-size distribution. In addition, the presence of graphene can magnify the contact of interparticle and increase electronic/ionic conductivities, which is great

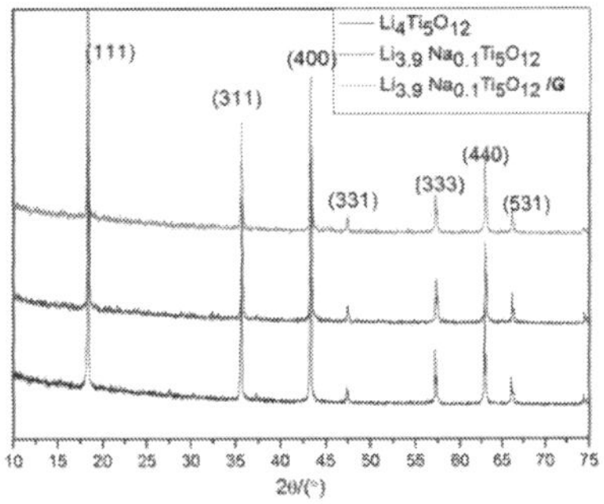

Fig.6 XRD patterns of the $Li_4Ti_5O_{12}$, $Li_{3.9}Na_{0.1}Ti_5O_{12}$ and $Li_{3.9}Na_{0.1}Ti_5O_{12}$/G samples.

importance in improving electrochemical performance. The following graph (Fig.8) shows the initial charge/discharge curves of the $Li_{3.9}Na_{0.1}Ti_5O_{12}$/G sample at 0.2C, 0.5C, 1C, 2C rate, the electrochemical performance has greatly improved compared to the pristine and $Li_{3.9}Na_{0.1}Ti_5O_{12}$ samples. This composite exhibits lower potential separation and outstanding reversible capacity no matter what on low or high current rates, possibly due to the addition of graphene. $Li_{3.9}Na_{0.1}Ti_5O_{12}$/G composites with a 3wt% of GO have a remarkable reversible capacity of 197.4mAh/g at 0.2C rate, significantly higher than the pristine electrode （167.4mAh/g）. The specific capacity decreases

Fig.7 FE-SEM images of $Li_{3.9}Na_{0.1}Ti_5O_{12}$/G composites (a,b) at different magnification

as the charge/discharge current rate increases, but even at 0.5C, 1C, 2C current rate, it still can present a discharge capacities of 175.5, 158.8 and 145.3mAh/g, respectively. Furthermore, the cycle performance is also improved due to the introduced of graphene, as described in Fig.9. The irreversible capacity has a large loss after the first cycle. However, it becomes more and more stable after several cycles, with the trend of cycle performance consistent with the pristine sample. The initial specific discharge capacity of the $Li_{3.9}Na_{0.1}Ti_5O_{12}$/G electrode is 158.8mAh/g at 1C rate, and maintains at 146.3mAh/g even after 40 cycles, with the reversible capacity retention is 92.13%, higher than the $Li_{3.9}Na_{0.1}Ti_5O_{12}$ sample at 87.73%. The results indicate that the cycling performance and capacity have greatly improved due to addition of graphene. Graphene with remarkable specific surface area and extraordinary electronic conductivity can greatly enlarge the contact of interparticles in the composite and change the intrinsic poor electronic and ionic conductivities of $Li_4Ti_5O_{12}$. In addition, the addition of graphene may effectively create numerous new Li^+ storage sites. This may be the reason why the $Li_{3.9}Na_{0.1}Ti_5O_{12}$/G composite has such excellent long-term cycling performance and impressive capacity.

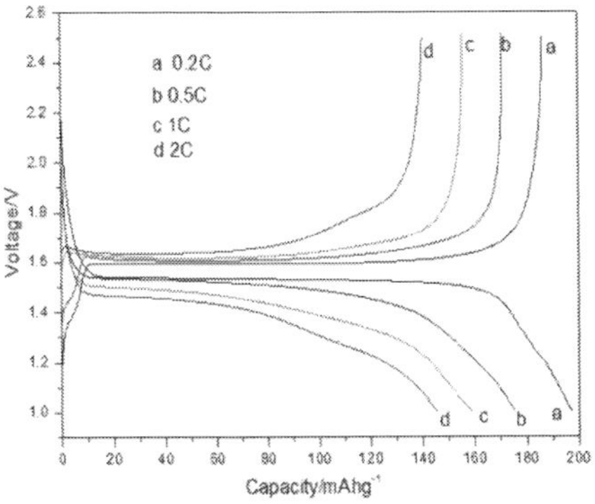

Fig.8 Initial charge/discharge curves of $Li_{3.9}Na_{0.1}Ti_5O_{12}$/G samples at 0.2C, 0.5C, 1C, 2C rates

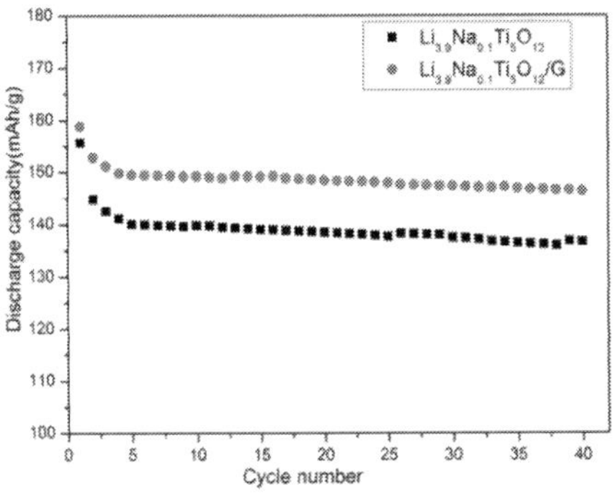

Fig.9 Cycling performance of the $Li_{3.9}Na_{0.1}Ti_5O_{12}$ and $Li_{3.9}Na_{0.1}Ti_5O_{12}$/G samples at 1 C rate

4. Conclusions

$Li_{4-x}Na_xTi_5O_{12}$ ($0\leq x\leq 0.15$) samples were successfully synthesized by a facile hydrothermal method. The results show that doping with Na^+ ions do not alter the structure of $Li_4Ti_5O_{12}$, and the resulting samples are well crystallized with a particle size in the range of 300-600nm. The Na-doped samples have higher specific capacities and excellent cycling stability, with the $Li_{3.9}Na_{0.1}Ti_5O_{12}$ electrode showing particularly good electrochemical properties among the four samples. The initial specific capacity of $Li_{3.9}Na_{0.1}Ti_5O_{12}$ electrode is 169.0mAh/g at 0.5C, and maintains at 150.4mAh/g even after 40 cycles. Finally, graphene was introduced to the composite by means of a simple solvothermal reduction method with ethanol as the solvent. This approach does not produce poisonous materials and has high efficiency. The initial specific discharge capacity of the $Li_{3.9}Na_{0.1}Ti_5O_{12}$/G electrode is 158.8mAh/g at 1C rate, and maintains at 146.3mAh/g even after 40cycles. The experiment results that the $Li_{3.9}Na_{0.1}Ti_5O_{12}$/G (3wt%) composites exhibit extraordinary electronic conductivity, faster lithium ion diffusivity, high-rate capability and long-term cycling stability, which makes it more practical in applications for lithium ion batteries.

References

[1] A. Khalll, K. Ryojik, T. Ynhua, Rechargeable lithium batteries and beyond: Progress, challenges, and future directions, MRS BULLETIN, 39 (2014).

[2] J.M. Tarascon, M. Armand, Issues and challenges facing rechargeable lithium batteries, Nature, 414 (2001).

[3] K. Ariyoshi, R. Yamato, T. Ohzuku, Zero-strain insertion mechanism of $Li[Li_{1/3}Ti_{5/3}]O_4$ for advanced lithium ion (shuttlecock) batteries, Electrocimica Acta, 51 (2005).

[4] K. Zaghib, M. Simoneau, M. Armand, M. Gauthier, Electrochemical study of as negative electrode for Li-ion polymer rechargeable batteries, J. Power Source, 81-82 (1999).

[5] T.F. Yi, Y. Xie, Y.R. Zhu, R.S. Zhu, H.Y. Shen, Structural and thermodynamic stability of $Li_4Ti_5O_{12}$ anode material for lithium-ion battery, J. Power Source, 222 (2013).

[6] S. Kim, S.H. Fang, Z.X. Zhang, J.Z. Chen, L. Yang, E.P.H. James, A. Deb, The electrochemical and local structural analysis of the mesoporous $Li_4Ti_5O_{12}$ anode, J. Power Source, 268 (2014).

[7] J.Z. Chen, L. Yang, S.H. Fang, Y.F. Tang, Synthesis of sawtooth-like $Li_4Ti_5O_{12}$ nanosheets as anode materials for Li-ion batteries, Electrochimica Acta, 55 (2010).

[8] Q. Zhou, L. Liu, J.L. Tan, Z.C. Yan, Z.F. Huang, X.Y. Wang, Synthesis of lithium titanate nanorods as anode materials for lithium and sodium ion batteries with superior electrochemical performance, J. Power Source, 283 (2015) .

[9] L. Sun, J.P. Wang, K.L. Jiang, S.S. Fan, Mesoporous $Li_4Ti_5O_{12}$ nanoclusters as high performance negative electrodes for lithium ion batteries, J. Power Source 248 (2014).

[10] H.H. Xu, X.L. Hu, Y.M. Sun, W. Luo, C.J. Chen, Y. Liu, Y.H. Huang, Highly porous $Li_4Ti_5O_{12}$/C nanofibers for

ultrafast electrochemical energy storage, Nano Energy, 10 (2014).

[11] W. Fang, X.Q. Cheng, P.J. Zuo, Y.L. Ma, G.P. Yin, A facile strategy to prepare nano-crystalline $Li_4Ti_5O_{12}$/C anode material via polyvinyl alcohol as carbon source for high-rate rechargeable Li-ion batteries, Electrochimica Acta 93 (2013) .

[12] J. Gao, J.R. Ying, C.Y. Jiang, C.R. Wan, High-density spherical $Li_4Ti_5O_{12}$/C anode material with good rate capability for lithium ion batteries, Journal of Power Source, 166 (2007) .

[13] C.P. Han, Y.B. He, H.F. Li, B.H. Li, H.D. Du, X.Y. Qin, F.Y. Kang, Suppression of interfacial reactions between $Li_4Ti_5O_{12}$ electrode and electrolyte solution via zinc oxide coating, Electrochimica Acta, 157 (2015) .

[14] H.P. Liu, G.W. Wen, S.F. Bi, P. Gao, Enhanced rate performance of nanosized $Li_4Ti_5O_{12}$/graphene composites as anode material by a solide state-assembly method, Electrochimica Acta, 171(2015).

[15] W.N. Chen, H. Jiang, Y.J. Hu, Y.H. Dai, C.Z. Li, Mesoporous single crystals $Li_4Ti_5O_{12}$ grown on rGO as high-rate anode materials for lithium-ion batteries, ChenCommun, 50 (2014).

[16] K.P. Loh, Q.L. Bao, P. Kailian, J.X. Yang, The chemistry of graphene, Journal of Materials Chemistry, 20 (2010) .

[17] N. Zhu, W. Liu, M.Q. Xue, Z. Xie, D. Zhao, M.N. Zhang, J.T. Chen, T.B. Cao, Graphene as a conductive additive to enhance the high-rate capabilities of electrospum $Li_4Ti_5O_{12}$ for lithium-ion batteries, Electrochimica Acta, 55(2010) .

MATEC Web of Conferences **65**, 02009 (2016) DOI: 10.1051/matecconf/20166502009
ICNSCM 2016

Combustion synthesis and electrochemical properties of $LiNi_{1/3}Co_{1/3}Mn_{1/3}Br_xO_{2-x}$ and $LiNi_{1/3}Co_{1/3}Mn_{1/3}Br_xO_{2-x}$/graphene cathode material for Li-ion batteries

Jiping Zhu*, Jie Zhu, Yangyang Zhang, Juan Wang and Jiawei Yan
School of Materials Science and Engineering, Hefei University of Technology, Hefei, Anhui230009, P R China

Abstract: The layered $LiNi_{1/3}Co_{1/3}Mn_{1/3}Br_xO_{2-x}$ ($0 \leq x \leq 0.09$) cathode materials were prepared by a combustion method. The XRD results indicate that the Br-doped $LiNi_{1/3}Mn_{1/3}Co_{1/3}O_2$ has the same layered structure as the pristine $LiNi_{1/3}Mn_{1/3}Co_{1/3}O_2$. FE-SEM results indicate that the particle size distribution of samples is uniform. Electrochemical tests reveal that Br-doped samples exhibit higher discharge capacity and rate capability compared with the pristine, especially the $LiNi_{1/3}Co_{1/3}Mn_{1/3}Br_{0.05}O_{1.95}$ sample shows initial discharge capacity, which can reach to 175.4 and 166.4mAh/g at 0.5 and 1.0C, respectively. Finally, an electronically conducting 2D network of graphene was introduced into $LiNi_{1/3}Co_{1/3}Mn_{1/3}Br_{0.05}O_{1.95}$ cathode material. The electrochemical properties of the materials were investigated by charge-discharge tests and electrochemical impedance spectroscopy. The charge-discharge tests demonstrate that this sample has better cycle stability than $LiNi_{1/3}Co_{1/3}Mn_{1/3}Br_{0.05}O_{1.95}$ which can be attributed to the excellent electronic conductivity and stable chemical properties of graphene. The EIS results reveal that the graphene coated greatly decreases the resistance of lithium batteries, especially the charge transfer resistance which can be attributed to the significantly improved electronic conductivity.

Key words: $LiNi_{1/3}Mn_{1/3}Co_{1/3}O_2$, Br-doping, graphene, combustion synthesis.

1. Introduction

During the past few decades, the rechargeable lithium-ion batteries (LIBs) have been one of the most important energy storage systems owing to its high energy, high energy density and good cycle performance. $LiCoO_2$ was the most commercial cathode material for LIBs due to its excellent cycle performance and relative ease of synthesis, but its application is limited because of the high price and the toxicity of cobalt. Therefore, much effort has been committed to developing $LiNi_xCo_yMn_{1-y}O_2$ layered compounds, especially for $LiNi_{1/3}Mn_{1/3}Co_{1/3}O_2$ in terms of its low cost, high discharge capacity and high thermal stability. Unfortunately, poor rate capability and fast capacity fading indeed restrict its widely application [1-4].

In order to overcome these serious problems of $LiNi_{1/3}Mn_{1/3}Co_{1/3}O_2$ material, researchers have done a lot of work. Summary previous reports, three methods are often used to improve its electrochemical performance. (1) Exploring an effective way to synthesis nanoparticles, which can shorten the diffusion distance of lithium ions and greatly increase the contact area between electrodes and electrolytes [1,5-6]. Recently, Hua et al successfully developed a fast co-precipitation method for the large scale synthesis of nanoflower structure $LiNi_{1/3}Co_{1/3}Mn_{1/3}O_2$ with the discharge capacity up to $126mAhg^{-1}$ at 20C when voltage range of 2.7-4.3V [5]. (2)Doping, partial cationic replace for the transition metal ions Ni, Co and Mn or anionic substitution for oxygen which can improve electronic conductivity. Such as,

Foundation item: Projects (No. 21373074) supported by the national natural science foundations of China and (No. 11010202133) supported by Anhui province science and technology plan
Corresponding author: Ji-ping ZHU; Tel:18919665197; E-mail: jpzh@ustc.edu.cn

Al^{3+}[7], Na^{+} [8], Ce^{4+} [9], Ti^{4+}[10], F^{-}[11], Y^{3+}[12], etc. (3) Surface modification, coating layer can prevent dissolution of transition metals ions into the electrolyte and side reactions between electrodes and electrolyte, enhancing the rate performance and cycle performance. Such as, CaF_2[13], TiO_2[14], Al_2O_3[15], LiF[16] and SrF_2[15].

Graphene (G) has been intensively investigated due to its excellent electronic conductivity, stable chemical properties, superior mechanical properties and high specific surface area. Recently, graphene was applied in lithium ion batteries as an ideal additive component for constructing hybrid nanostructure of LIB electrodes [18-20]. Jiang et al successfully synthesize a composite material of graphene wrapped $Li_2MnO_3 \cdot LiMO_2$, (M=Mn, Ni, Co) cathode material. Compared to the pure $Li_2MnO_3 \cdot LiMO_2$, the graphene conducting framework can efficiently alleviate the polarization of pristine $Li_2MnO_3 \cdot LiMO_2$ material leading to an outstanding enhancement in cell performance and cycling stability [19].

This paper employs a simple combustion method to synthesize Br doping LNCM material. To the best of our knowledge, this is the first time that Br doping LNCM material by combustion method. This simple method could uniformly mix the metal ions and provide more heat for the reaction [6]. Compare with combustion method, traditional solid state method needs high calcination temperature and long reaction time resulting inhomogeneous or impurity with lower capacity and poor cycling performance. Co-precipitation method needs strict control the condition of reaction and quite long reaction time which is not conducive to industrial production. Sol-gel is similar to combustion method, but combustion method does not require adjustment PH [1, 8, 21-22]. Br^{-} ion substitution for O^{2-} ion may lead to an increased electronic/ionic conductivity and improved reversible specific capacity. In addition, graphene was introduced for the best Br content of the complex, which is utilize for excellent electron conductivity and stable chemical properties of graphene.

2. Experimental

2.1 Synthesis of $LiNi_{1/3}Co_{1/3}Mn_{1/3}O_2$ with Br-doping

$LiNi_{1/3}Co_{1/3}Mn_{1/3}O_2$ material was synthesized by combustion method using tartaric acid as combustion agent. Firstly, the stoichiometric amounts of lithium acetate (5% excess amount of lithium acetate was used to compensate Li loss during the calcination), nickel acetate, cobalt acetate and manganese acetate were dissolved in ethylene glycol. In addition, the molar ratio of the tartaric acid to metal ions was 1:1. Secondly, the mixed solution was under magnetic stirring at 25℃for 1h and then was heated at 150℃ for 6h. Thirdly, the obtained mixture was ground and then put into an Al_2O_3 crucible. The crucible was put in a muffle furnace preheated at 450℃ for 6 h in air to eliminate the organic composition and then was sintered at 850℃ for 10h. The sample heating rate of 4℃min^{-1} and then slowly cooled to room temperature.

In order to obtain the Br-doped samples, the same procedures were carried out, only at the beginning adding lithium bromide. $LiNi_{1/3}Co_{1/3}Mn_{1/3}Br_xO_{2-x}$ (x=0.01, 0.03, 0.05, 0.07, 0.09) samples were gained by adjusting the content of LiBr.

2.2 Synthesis of $LiNi_{1/3}Co_{1/3}Mn_{1/3}Br_{0.05}O_{1.95}$/G composite

Graphite oxide (GO) nanosheets were fabricated via a modified Hummers method using natural graphite powders as the raw material. 2.0g/L graphene oxide solution was prepared by graphite oxide powder dissolve into ethanol solution followed by ultrasonication for 0.5h. Meanwhile, the previous synthesis $LiNi_{1/3}Co_{1/3}Mn_{1/3}Br_{0.05}O_{1.95}$ was dissolved into ethanol solution and vigorously stirred for 30min. Then, the previously prepared GO solution was added into the $LiNi_{1/3}Co_{1/3}Mn_{1/3}Br_{0.05}O_{1.95}$ suspension with 3wt% content of GO. Subsequently, the mixed solution was transferred to a Teflon-lined autoclave at 120℃for 3h. As the end of reaction, the products were filtered and washed using distilled water and ethanol several times, then dried at 65℃in an oven.

2.3 Material characterization and electrochemical measurements

The crystalline structure of the samples was analyzed by an X-ray diffraction (XRD, X'Pert PRO MPD) measurements using Cu Kα radiation. The scan data was recorded in the 2θ range of 10–80°. The field-emission scanning electron microscope (FE-SEM, SU8020)

DOI: 10.1051/matecconf/20166502009

equipped with an energy dispersive spectroscope (EDS) was used to study the morphologies and elements distribution of samples. The powder was dispersed into ethanol and a drop of the suspension was transferred onto aluminum foil and dried for FE-SEM observations.

The electrochemical performance of the as-synthesized material was investigated in CR2032 coin cells. The coin cell contained a test electrode (cathode), a pure lithium foil (anode), a microporous polyethylene membrane, and an electrolyte solution. For the test electrode, a mixture of active material, acetylene black, and polyvinylidene fluoride (PVDF) binder at a weight ratio of 8:1:1 was dispersed in a N-methylpyrrolidone (NMP) solution. The slurry was paste on an aluminum foil, and dried at 120℃ for 12h in a vacuum oven. The pure lithium foil was used as a reference and counter electrode, the microporous polyethylene membrane (Celgard 2400) was used as a separator, and the electrolyte was 1 M $LiPF_6$ in a 1:1 (v/v) mixture of ethylene carbonate (EC) and diethyl carbonate (DEC). The cells were assembled in an argon-filled glove box, afterwards aged for 12 h before electrochemical test. The electrochemical impedance spectroscopy EIS) was performed on CHI660D electrochemical workstation .The frequency range was from 0.01Hz to 100k Hz and amplitude voltage was 5 mV.

3. Results and discussion

3.1 The effect of Br-doping on crystal structure, morphology and electrochemical performance

Fig. 1 illustrated the XRD patterns of the pristine and Br-doping $LiNi_{1/3}Co_{1/3}Mn_{1/3}O_2$ cathode materials. All the observed diffraction peaks are well indexed in the R-3m space with α-$NaFeO_2$ layered structure. No impurity peaks were observed, which indicated that the Br-doping does not change on the crystalline structure. The peak intensity ratio of (003)/(104) is sensitive to the degree of cation mixing in the layered compounds, and the larger I(003)/I(104) value, the less cation mixing between Li and Ni ions. A higher value (>1.2) of I(003)/I(104) is generally believed that have less cation mixing and better hexagonal structure in the layered compounds [3]. The I(003)/I(104)values of all samples are larger than 1.2, it was suggested that less cation mixing takes place. In addition, the pairs of (006)/(102) and (108)/(110) are spilt, which is considered to be the characteristic of the layered structure. It is clear that split of the (006)/(102) and (108)/(110) peaks when the doping amount of Br is less than 5%. When the content of Br doping continues to increase, the peaks split are not obvious. This phenomenon may be due to the fact that excess Br-doping decreasing crystallization of the material and it is not conducive to form stable layered structure.

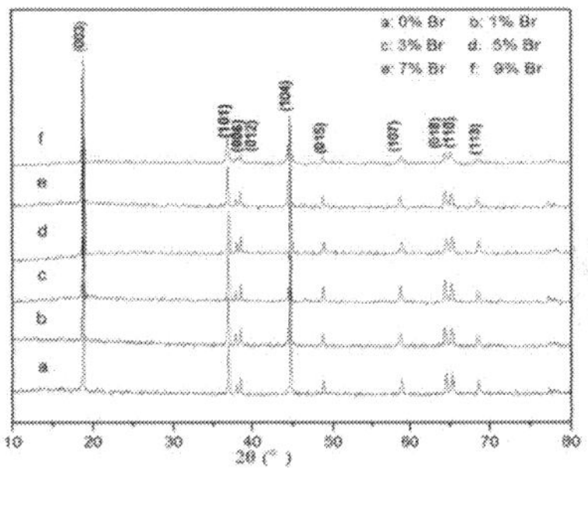

Fig.1 XRD patterns of $LiNi_{1/3}Co_{1/3}Mn_{1/3}Br_xO_{2-x}$ (x=0, 0.01, 0.03, 0.05, 0.07, 0.09)

FE-SEM images of the pristine and $LiNi_{1/3}Co_{1/3}Mn_{1/3}Br_{0.05}O_{1.95}$ are shown in Fig. 2. The pristine and Br-doping are composed of small particles with smooth surface and particle size in the range of 300nm-500nm. Compared with the pristine, the $LiNi_{1/3}Co_{1/3}Mn_{1/3}Br_{0.05}O_{1.95}$ has more homogeneous particle-size distributions, which is beneficial to intercalation and de-intercalation of lithium ions. It may bring about good electrochemical performance. Furthermore, in order to prove the homogeneity of the $LiNi_{1/3}Co_{1/3}Mn_{1/3}O_2$, the EDS results are shown in Fig.3. The results shows that transition metal elements and oxygen are homogeneously distribution on the surface of particle, and the results agreed with the XRD results, further demonstrating that the final product was $LiNi_{1/3}Co_{1/3}Mn_{1/3}O_2$.

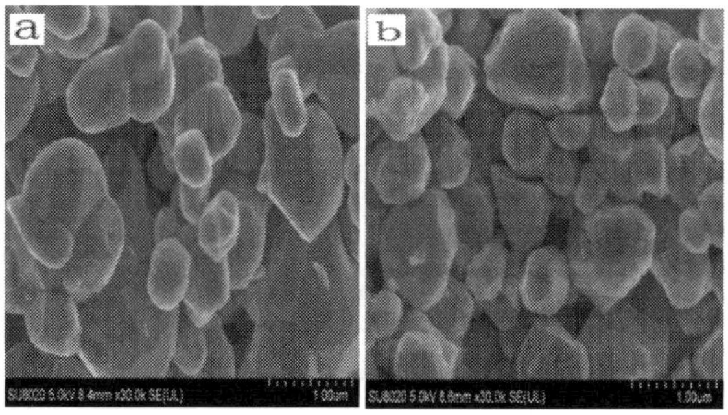

Fig.2 FE-SEM images of $LiNi_{1/3}Co_{1/3}Mn_{1/3}Br_xO_{2-x}$ (x=0, 0.03) (a) x=0, (b) x=0.03

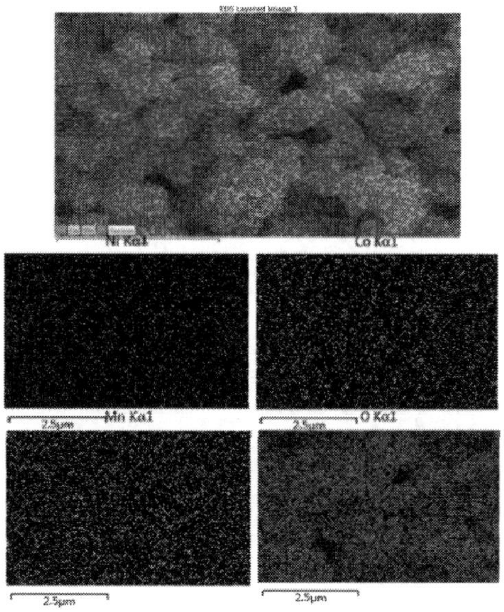

Fig.3 EDS element mapping of $LiNi_{1/3}Co_{1/3}Mn_{1/3}O_2$

The initial charge–discharge curves of the prepared $LiNi_{1/3}Co_{1/3}Mn_{1/3}Br_xO_{2-x}$ (x=0, 0.01, 0.03, 0.05, 0.07, 0.09) shown in Fig. 4 between 2.5 and 4.3 V at 0.2 C. Charge tests were performed using a constant current follow by constant voltage charge, which means charging continued at 4.3V at a constant current (0.2C), and then kept at the voltage of 4.3V until the current dropped to less than 0.02C, and then discharged to 2.5V cut-off potential in constant current. As can be seen from Fig. 4, the initial specific discharge capacities of $LiNi_{1/3}Co_{1/3}Mn_{1/3}Br_xO_{2-x}$ (x= 0, 0.01, 0.03, 0.05, 0.07, 0.09) are 157.4, 169.9, 173.7, 179.6, 156.1and 147.8mAh/g, respectively. The initial discharge capacity increased dramatically when the Br content of x is less than or equal to 0.05. The likely reason was the O^{2-} ion (0.14nm) substituted by Br^- ion (0.196nm) lead to the lattice parameter increase, which in turn may lead to an increased electronic/ionic conductivity and improved reversible specific capacity. The discharge capacity decreased when the content of Br doping continues to increase. It was suggested the excess Br-doping caused the structural instability and deteriorated of the electrochemical performance.

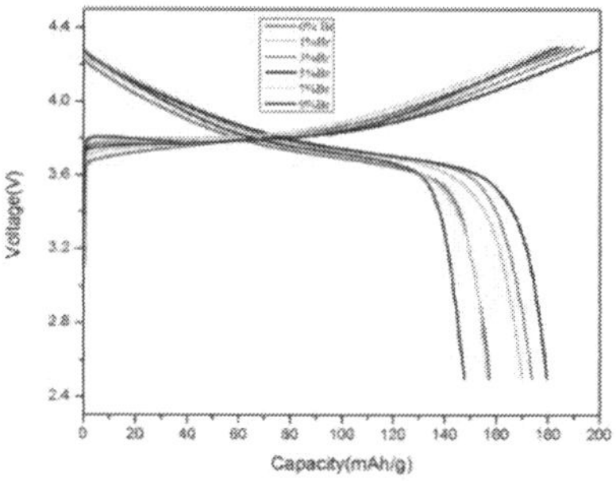

Fig.4 Initial charge/discharge curves of $LiNi_{1/3}Co_{1/3}Mn_{1/3}Br_xO_{2-x}$ (x=0, 0.01, 0.03, 0.05, 0.07, 0.09) samples at 0.2C

Fig. 5 showed the rate capacity of the pristine and $LiNi_{1/3}Co_{1/3}Mn_{1/3}Br_{0.05}O_{1.95}$ samples. The cells were charged to 4.3V at the current 0.1C, and then discharged to 2.5V at 0.2, 0.5, 1.0 and 2.0 C, respectively. Obviously, with the increase of current density, the discharge capacity of the two samples reduced gradually. Moreover, the $LiNi_{1/3}Co_{1/3}Mn_{1/3}Br_{0.05}O_{1.95}$ sample delivers higher capacity than the pristine at the same currents. The $LiNi_{1/3}Co_{1/3}Mn_{1/3}Br_{0.05}O_{1.95}$ sample shows a high discharge capacity of 175.4, 166.4 and 153.9mAh/g at 0.5, 1.0 and 2.0C, remaining 97.6%, 92.6%and 87.7% of the capacity of 179.6mAh/g at the 0.2 C. The improved rate capacity of the $LiNi_{1/3}Co_{1/3}Mn_{1/3}Br_{0.05}O_{1.95}$ sample can be partially ascribed to the increased electronic/ionic conductivity due to Br-doping.

 DOI: 10.1051/matecconf/20166502009

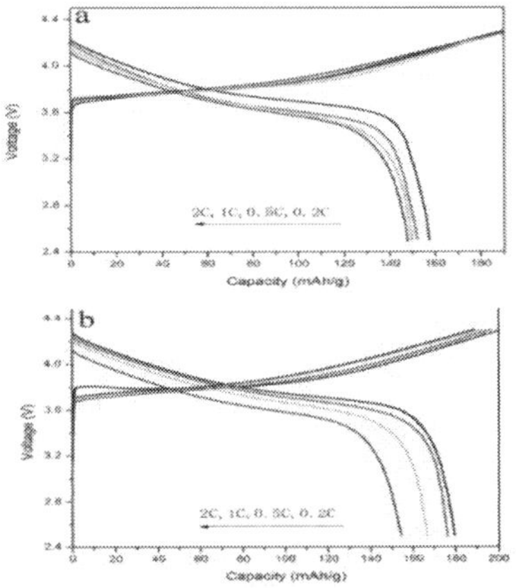

Fig.5 Rate capacity of (a) $LiNi_{1/3}Co_{1/3}Mn_{1/3}O_2$ and (b) $LiNi_{1/3}Co_{1/3}Mn_{1/3}Br_{0.05}O_{1.95}$ samples

Cycling performance of the pristine and $LiNi_{1/3}Co_{1/3}Mn_{1/3}Br_{0.05}O_{1.95}$ are shown in Fig. 6. In both cases, the specific capacities decreased gradually during cycles, which may be attributed to the corrosion of the electrode by the electrolytes. After 50 charge/discharge cycles, the specific discharge capacity of $LiNi_{1/3}Co_{1/3}Mn_{1/3}Br_{0.05}O_{1.95}$ sample is 133.1mAh/g, however, its retention rate is lower than the pristine, which the pristine retained around 78.8% of the initial discharge capacity. The $LiNi_{1/3}Co_{1/3}Mn_{1/3}Br_{0.05}O_{1.95}$ sample suffered a severe capacity fading, which may be lead to transition ions easily dissolved in the electrolyte caused by Br doping.

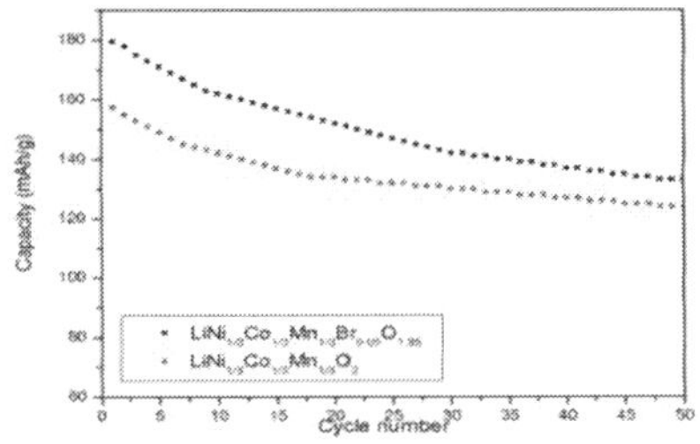

Fig.6 Cycling performance of $LiNi_{1/3}Co_{1/3}Mn_{1/3}O_2$ and $LiNi_{1/3}Co_{1/3}Mn_{1/3}Br_{0.05}O_{1.95}$ samples at 0.2C

3.2 Crystal structure, morphology and electrochemical performance of $LiNi_{1/3}Co_{1/3}Mn_{1/3}Br_{0.05}O_{1.95}$/G sample

Recent advances show that graphene can serve as a perfect 2D support on the edges and surface for anchoring metal or metal oxide nanoparticles [17-18, 23]. GO was fabricated via a modified Hummers method contain many hydrophilic oxygen-containing functional groups, such as –COOH, -OH, C=O and epoxy groups, which enables GO to be well dispersed [25]. Solvothermal method is considered to be an effective and practical method for producing graphene/metal oxide composite materials. In order to further prove solvothermal method is involved in the reduction of GO to G. XRD patterns of the GO and G samples are show in Fig.7 (a). The XRD patterns of the $LiNi_{1/3}Co_{1/3}Mn_{1/3}Br_{0.05}O_{1.95}$ and $LiNi_{1/3}Co_{1/3}Mn_{1/3}Br_{0.05}O_{1.95}$/G are depicted in Fig.7(b) . All diffraction peaks can be indexed in the R-3m space with α-$NaFeO_2$ structure. All samples showing a typical hexagonal layered structure because the (006)/(102) and (108)/(110) peaks are well separated. In addition, graphene characteristic XRD peaks not being detected in the XRD pattern, this may be due to the low content of graphene.

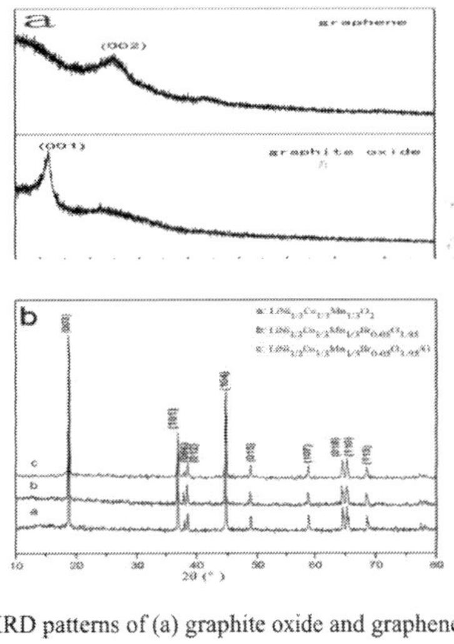

Fig.7 XRD patterns of (a) graphite oxide and graphene (b)$LiNi_{1/3}Co_{1/3}Mn_{1/3}O_2$, $LiNi_{1/3}Co_{1/3}Mn_{1/3}Br_{0.05}O_{1.95}$, $LiNi_{1/3}Co_{1/3}Mn_{1/3}Br_{0.05}O_{1.95}$/G

Fig.8 shows the FE-SEM images of the $LiNi_{1/3}Co_{1/3}Mn_{1/3}Br_{0.05}O_{1.95}$/*G* composites at different magnifications. Graphene sheets are uniformly distributed among the $LiNi_{1/3}Co_{1/3}Mn_{1/3}Br_{0.05}O_{1.95}$ particles, giving a better connection between adjacent $LiNi_{1/3}Co_{1/3}Mn_{1/3}Br_{0.05}O_{1.95}$ particles, as well as preventing them from further aggregation. In addition, the presence of graphene may be inhibit the dissolution of metal ions and increase electronic/ionic conductivities, which is of great importance in improving electrochemical performance.

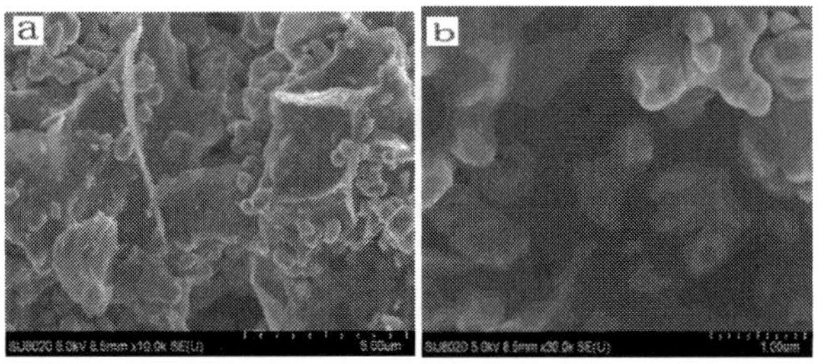

Fig.8 FE-SEM images of $LiNi_{1/3}Co_{1/3}Mn_{1/3}Br_{0.05}O_{1.95}/G$

The initial charge–discharge curves of the $LiNi_{1/3}Co_{1/3}Mn_{1/3}Br_{0.05}O_{1.95}$ and $LiNi_{1/3}Co_{1/3}Mn_{1/3}Br_{0.5}O_{1.95}/G$ are shown in Fig. 9. Both samples clearly displayed a potential plateau around 3.6–3.8 V during the charge-discharge processes, which in agreement with typical layer structured $LiNi_{1/3}Co_{1/3}Mn_{1/3}O_2$ reported by many researchers [4, 20]. The specific discharge capacities of the $LiNi_{1/3}Co_{1/3}Mn_{1/3}Br_{0.05}O_{1.95}$ and $LiNi_{1/3}Co_{1/3}Mn_{1/3}Br_{0.5}O_{1.95}/G$ are 179.6 and 181.3 mAh/g at 0.2C, respectively. Both samples of the first discharge specific capacity is very similar, but their cycle performance difference, as shown in Fig. 10. It is clear that the discharge capacity of the graphene coated reach to 145mAh/g after 50 cycles, higher than that of the $LiNi_{1/3}Co_{1/3}Mn_{1/3}Br_{0.05}O_{1.95}$ sample at 133.1mAh/g. One possible reason is that the graphene coated stabilizes the surface structure of the layered oxide, thereby reduced the dissolution of metal ions.

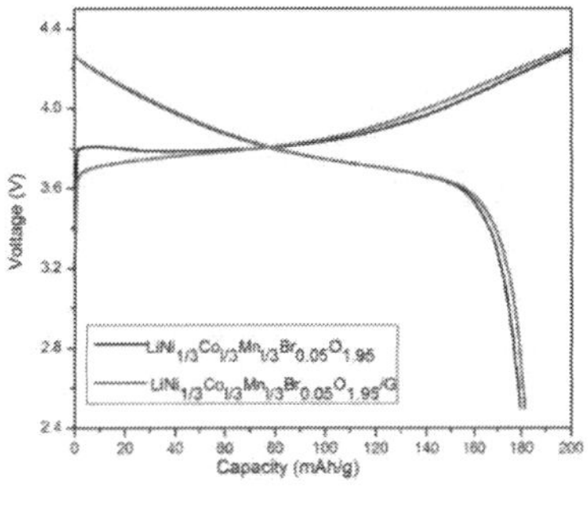

Fig.9 Initial charge/discharge curves of $LiNi_{1/3}Co_{1/3}Mn_{1/3}Br_{0.5}O_{1.95}$ and $LiNi_{1/3}Co_{1/3}Mn_{1/3}Br_{0.5}O_{1.95}$ /G samples at 0.2C

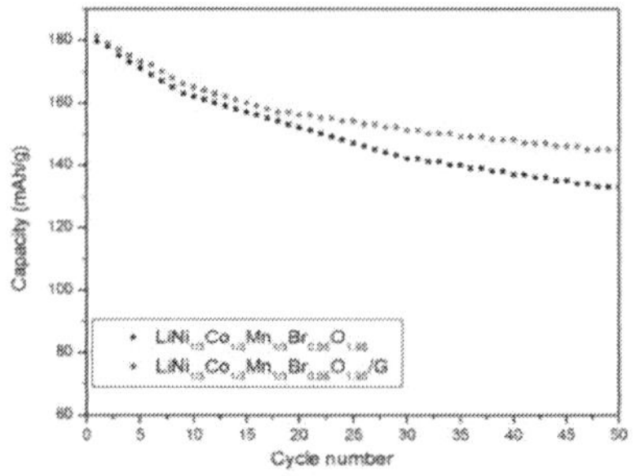

Fig.10 Cycling performance of $LiNi_{1/3}Co_{1/3}Mn_{1/3}Br_{0.05}O_{1.95}$ and $LiNi_{1/3}Co_{1/3}Mn_{1/3}Br_{0.5}O_{1.95}/G$ samples at 0.2C

The electrochemical impedance spectra (EIS) of the $LiNi_{1/3}Co_{1/3}Mn_{1/3}Br_{0.05}O_{1.95}$ and $LiNi_{1/3}Co_{1/3}Mn_{1/3}Br_{0.5}O_{1.95}/G$ electrodes are shown Fig. 11. The impedance spectra exhibit two semicircles in the high and intermediate-frequency ranges and followed by a straight slopping line in the low frequency region. The high frequency semicircle is related the Li^+ ion migration resistance (R_{sei}) through the solid electrolyte interface (SEI) film that forms on the surface of the cathode materials; the intermediate frequency semicircle is attributed to the charge transfer resistance (R_{ct}) in the cathode/electrolyte interface; the low frequency tail is attribute to Warburg impedance that is associated with Li^+ diffusion through the cathode [25-28]. From the Fig.11, it could also see that the first semicircle has virtually no change of these two samples, which means the solid electrolyte interface layer formed on the surface of the electrode has no great difference. However, the R_{ct} of $LiNi_{1/3}Co_{1/3}Mn_{1/3}Br_{0.05}O_{1.95}/G$ sample is significantly smaller than that of $LiNi_{1/3}Co_{1/3}Mn_{1/3}Br_{0.05}O_{1.95}$. One interpretation is that it is the graphene coated improved electronic conductivity and stabilizes the surface structure of the layered oxide, thus the dissolution of metal ions is reduced. Small impedance is conducive to the intercalation and de-intercalation of lithium-ions during the charge and discharge process. Hence, the EIS test also can confirm the improvement of electrochemical properties.

DOI: 10.1051/matecconf/20166502009

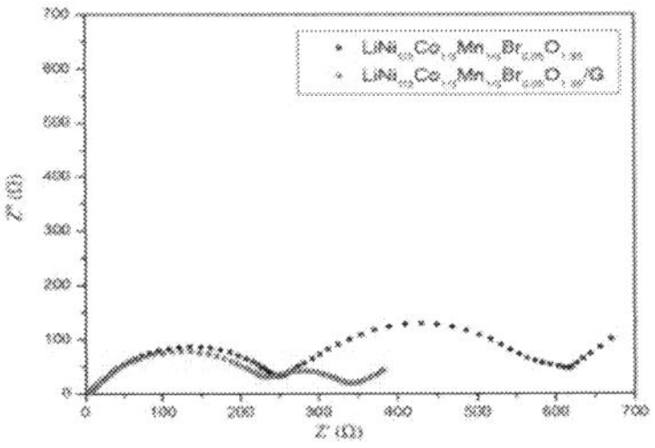

Fig.11 Electrochemical impedance spectra of $LiNi_{1/3}Co_{1/3}Mn_{1/3}Br_{0.05}O_{1.95}$ and $LiNi_{1/3}Co_{1/3}Mn_{1/3}Br_{0.5}O_{1.95}$ /G samples

4. Conclusions

Layered-structured $LiNi_{1/3}Co_{1/3}Mn_{1/3}Br_xO_{2-x}$ ($0 \leq x \leq 0.08$) materials were successful synthesized by combustion method. The structure, morphology and electrochemical properties of samples were investigated in detail. The results show that Br-doping does not alter the structure of $LiNi_{1/3}Mn_{1/3}Co_{1/3}O_2$, and exhibits higher discharge capacity, better cycle performance and rate capability. Particularly the $LiNi_{1/3}Co_{1/3}Mn_{1/3}Br_{0.05}O_{1.95}$ sample shows a discharge capacity of 179.6, 175.4, 166.4 and 153.9mAh/g at 0.2, 0.5, 1.0 and 2.0C, respectively. Finally, graphene was introduced to the composite through a simple solvothermal method. The cycle performance of $LiNi_{1/3}Co_{1/3}Mn_{1/3}Br_{0.05}O_{1.95}$/G (3wt%) composites are much better than $LiNi_{1/3}Co_{1/3}Mn_{1/3}Br_{0.05}O_{1.95}$. EIS results reveal that graphene modification improved electronic conductivity and reduce the charge transfer impedance of material.

References

[1] F.Wu, M. Wang, Y.f. Su, A novel method for synthesis of layered $LiNi_{1/3}Mn_{1/3}Co_{1/3}O_2$ as cathode material for lithium-ion battery, J. Power Sources, **195** (2010)

[2] C.X. Ding, Y.C. Bai, X.Y. Feng, Improvement of electrochemical properties of layered $LiNi_{1/3}Co_{1/3}Mn_{1/3}O_2$ positive electrode material by zirconium doping, Solid State Ionics, **189** (2011)

[3] X.Z Liu, P. He, H.Q Li, Improvement of electrochemical properties of $LiNi_{1/3}Co_{1/3}Mn_{1/3}O_2$ by coating with V_2O_5 layer, J.Alloys Compd., **552** (2013)

[4] P. Robert Ilango, T. Subburaj, K. Prasanna, Physical and electrochemical performance of $LiNi_{1/3}Co_{1/3}Mn_{1/3}O_2$ cathodes coated by Sb_2O_3 using a sol-gel process, Mater. Chem. Phys., **158** (2015)

[5] W.B. Hua, X.D. Guo, Z. Zheng, Uncovering a facile large-scale synthesis of $LiNi_{1/3}Co_{1/3}Mn_{1/3}O_2$ nanoflowers for high power lithium-ion batteries, J. Power Sources, **275** (2015)

[6] J.L. Zheng, W. Zhou, Y.R. Ma, Combustion synthesis of $LiNi_{1/3}Co_{1/3}Mn_{1/3}O_2$ powders with enhanced electrochemical performance in LIBs, J.Alloys Compd., **635** (2015)

[7] Y.H.Ding, P. Zhang, Z.L. Long, Morphology and electrochemical properties of Al doped $LiNi_{1/3}Co_{1/3}Mn_{1/3}O_2$ nanofibers prepared by electrospinning, J.Alloys Compd., **487** (2009)

[8] C.X. Gong, W.X. Lv, L.M. Qu, Syntheses and electrochemical properties of layered $Li_{0.95}Na_{0.05}Ni_{1/3}Co_{1/3}Mn_{1/3}O_2$ and $LiNi_{1/3}Co_{1/3}Mn_{1/3}O_2$, J. Power Sources, **247** (2014)

[9] S.K.Zhong, Y.Wang, J.Q.Liu, Synthesis and electrochemical properties of Ce-doped $LiNi_{1/3}Mn_{1/3}Co_{1/3}O_2$ cathode material for Li-ion batteries,J. rare earths, **299**(2011).

[10] G.Y Li, Z.L. Huang, Z.L. Zuo, Understanding the trace Ti surface doping on promoting the low temperature performance of $LiNi_{1/3}Co_{1/3}Mn_{1/3}O_2$ cathode, J. Power Sources, **281** (2015)

[11] Masaya Kageyama, Decheng Li, Koichi Kobayakawa, Structural and electrochemical properties of $LiNi_{1/3}Mn_{1/3}Co_{1/3}O_{2-x}F_x$ prepared by solid state reaction, J. Power Sources, **157** (2006)

[12] Y.W Li, Y.X. Li, S.K. Zhong, Synthesis and Electrochemical Properties of Y-Doped $LiNi_{1/3}Co_{1/3}Mn_{1/3}O_2$ Cathode Materials for Li-Ion Battery, Integrated Ferroelectrics, **127**(2011)

[13] Ke Xua, Zhifu Jie a,b, Ronghua Li, Synthesis and electrochemical properties of CaF_2-coated for long-cycling $Li[Mn_{1/3}Co_{1/3}Ni_{1/3}]O_2$ cathode materials, Electrochimica Acta, **60** (2012)

[14] F.Wu, M. Wang, Y.f. Su, Effect of TiO_2-coating on the electrochemical performances of $LiNi_{1/3}Co_{1/3}Mn_{1/3}O_2$, J. Power Sources, **191** (2009)

[15] X.L Li, W.X. He, L. Chen, Hydrothermal synthesis and electrochemical performance studies of

Al_2O_3-coated $LiNi_{1/3}Co_{1/3}Mn_{1/3}O_2$ for lithium-ion batteries, Ionics, **20** (2014)

[16] S.J. Shi, J.P. Tu, Y.Y. Tang, Enhanced electrochemical performance of LiF-modified $LiNi_{1/3}Co_{1/3}Mn_{1/3}O_2$ cathode materials for Li-ion batteries, J. Power Sources, **225** (2013)

[17] J.G. Li, L. Wang, Q. Zhang, Electrochemical performance of SrF_2-coated $LiNi_{1/3}Co_{1/3}Mn_{1/3}O_2$ cathode materials for Li-ion batteries, J. Power Sources, **190** (2009)

[18] Z.S. Wu, G.M. Zhou, L.C. Yin, Graphene/metal oxide composite electrode materials for energy storage, Nano Energy, **1** (2012)

[19] K.C. Jiang, X.L. Wu, Y.X. Yin, Superior Hybrid Cathode Material Containing Lithium-Excess Layered Material and Graphene for Lithium-Ion Batteries, ACS Applied Materials & Interfaces, **49** (2012)

[20] K.C. Jiang, S Xin, J.S. Lee, Improved kinetics of $LiNi_{1/3}Mn_{1/3}Co_{1/3}O_2$ cathode material through reduced graphene oxide networks, Phys. Chem. Chem. Phys., **14** (2012)

[21] C. Deng, S. Zhang, L. Ma, Effects of precipitator on the morphological, structural and electrochemical characteristics of $Li[Ni_{1/3}Co_{1/3}Mn_{1/3}]O_2$ prepared via carbonate coprecipitation, J.Alloys Compd., **509** (2011)

[22] P. Gao, G. Yang, H.D. Liu, Lithium diffusion behavior and improved high rate capacity of $LiNi_{1/3}Co_{1/3}Mn_{1/3}O_2$ as cathode material for lithium batteries, Solid State Ionics,**207** (2012)

[23] I.V. Lightcap, T.H. Kosel, P.V. Kamat, Anchoring Semiconductor and Metal Nanoparticles on a Two-Dimensional Catalyst Mat.Storing and Shuttling Electrons with Reduced Graphene Oxide, Nano Letters, **10** (2010)

[24] Gints Kucinskis, Gunars Bajars, Janis Kleperis,Graphene in lithium ion battery cathode materials: A review, J. Power Sources, **240** (2013)

[25] F.Wu, M. Wang, Y.f. Su, A novel layered material of $LiNi_{0.32}Mn_{0.33}Co_{0.33}Al_{0.01}O_2$ for advanced lithium-ion batteries, J. Power Sources **195** (2010)

[26] Y.Y. Hu, Y.K. Zhou, J. Wang, Preparation and characterization of macroporous $LiNi_{1/3}Co_{1/3}Mn_{1/3}O_2$ using carbon sphere as template, Mater. Chem. Phys. **129** (2011)

[27] G.T.-K. Fey, P. Muralidharan, C.-Z. Lu, Y.-D. Cho, Enhanced electrochemical performance and thermal stability of La_2O_3-coated $LiCoO_2$, Electrochim Acta, **51** (2006)

[28] Z.R. Zhang, H.S. Liu, Z.L Gong, Electrochemical performance and spectroscopic characterization of TiO_2-coated $LiNi_{0.8}Co_{0.2}O_2$ cathode materials, J. Power Sources, **129** (2004)

MATEC Web of Conferences **65**, 02010 (2016) DOI: 10.1051/matecconf/20166502010
ICNSCM 2016

Synthesis of Pd@Pt Core-shell Nanoparticles based on Photochemical Seed Growth Method and Co-reduction Method and the Electrocatalytic Performance

Shanshan Li[1] Yingnan Dong[1,a] Xiaoguo Bi[1] Xudong Liu[1] and Wei Niu[1]

[1] *Shenyang Institute of Engineering, Shenyang 110136, China*

Abstract. A series of Pd@Pt nanoparticles were synthesized based on electrochemical seed growth method and co-reduction method in polyethylene-glycol and acetone solution system. The TEM/HR-TEM and XPS characterization proved that the prepared composite nanoparticles present core-shell structure and analyzed the chemical state of the particles. The electrocatalytic performance of Pd@Pt particles was studied by using the electrochemical workstation. The results showed that the Pd@Pt/C catalyst of different molar ratios of Pd to Pt exhibited preferable catalytic activity and stability for the methanol catalytic oxidation reaction. Among which, the Pd@Pt nanoparticles (Pd:Pt=1:1) prepared by co-reduction method, presented highest catalytic activity, which is 2 times higher than that of Pt/C catalyst. The high catalytic activity produced by the core-shell structure was briefly discussed.

1 Introduction

The shortage of fossil fuel, and the atmosphere pollution and greenhouse effect caused by the usage of fossil fuel are problems need to be solved urgently in energy and environmental fields. Low-temperature fuel cells, including direct methanol fuel cell (DMFC) and proton exchange membrane fuel cell (PEMFC), exhibit an energy conversion rate as high as 60~80% and the actual service efficiency is 2~3 times of that of common internal combustion engine. It is a green, energy saving and high efficient alternative energy which is applied widely. The research and development of low-temperature fuel cell have always focused on the search of electrode catalyst with high performance and the mechanism for improving the catalytic performance[1-3].

Owing to the cheaper price of Pd than Pt and the widely applied catalytic activity, Pd has been introduced in the electrocatalysis of fuel cell recently. Y. H. Cho et al[4] synthesized $PtPd_{alloy}/C$ PEMFC anode catalyst, which presented similar catalytic performance with Pt/C. But the atomic ratio of Pt to Pd was merely 1:19. Regarding core-shell catalyst, S.P. Jiang et al[5] synthesized Pd@Pt core-shell nano-wire by adopting hard template electrochemical deposition method. The composite nano-material has large specific surface area and shows well catalytic activity for the methanol catalytic oxidation reaction (MOR). K. Sasaki et al[6] conducted experiments in which single layer of Cu atoms were deposited on the surface of Pd particles of 4 nm in diameter; then the Cu atoms were replaced by Pt atoms (underpotential deposition method). Finally, they synthesized Pd@Pt nanoparticles with Pt shell of single atom layer. The current density of Pt in unit mass for MOR is about 2.9 times of that of Pt/C. Generally, the research on Pd-Pt catalyst is still in a fledging period; while the catalyst has shown its favourable cost advantage and application prospect.

Two novel methods for synthesizing Pd@Pt core-shell composite nanoparticles were presented based on the photochemistry. In the polyethylene-glycol (PEG)-acetone solution system, the nanoparticles were synthesized by ultraviolet irradiation to reduce the mixture of Pd (II) and Pt (IV) ions simultaneously (co-reduction method); regarding the seed growth method, Pd was applied as the crystal seed, and the Pt (IV) ions were irradiated by ultraviolet light to reduce and deposit on the surface of Pd seed. The characterization applying UV-Vis absorption spectra, TEM/HR-TEM and XPS verified that Pd@Pt core-shell composite nanoparticles with uniform size and monodispersion were obtained by employing the both methods. Furthermore, the size of the composite particles and the thickness of the Pt shell can be efficiently adjusted by changing the molar ratio of Pd (II) or Pd seed to Pt (IV). The average diameter of the composite nanoparticles synthesized adopting co-reduction method is in the range of 4.1~2.9 nm; and that using the seed growth method is in the range of 5.3~7.1 nm.

Afterward, the obtained Pd@Pt nanoparticles were loaded on XC-72 carbon black, which was treated with $NaBH_4$, to systematically study the activity for MOR. The research showed that the Pd@Pt core-shell composite nanoparticles present high catalytic activity for

[a] Corresponding author: dyn4870@163.com

methanol. The authors considered that the high activity resulted from the promotion of Pd to Pt.

2 Experimental methods

2.1 Preparation of products

Co-reduction: Certain amounts of PEG-400, acetone and different molar ratios of H_2PdCl_4 and H_2PtCl_6 (the molar ratios of Pd to Pt elements are 4:1, 1:1 and 1:4, respectively, and the total molar amount of noble metal is 1.22×10^{-5} mol). Then water was used to get the constant volume. After being mixed sufficiently, the solution was transferred into a quartz container and then exposed 4 cm away under ultraviolet light. UV-Vis spectra were recorded as time passed by until the spectral curve was the same as the previous one, which indicated that the reduction finished. The single Pd and Pt nanoparticles were prepared similarly.

Seed growth method: Certain amounts of Pd colloidal solution (crystal seed) and H_2PtCl_6 solution were added in the quartz container (the molar ratios of Pd to Pt are 4:1, 1:1 and 1:4, respectively; the total molar amount of noble metal does not change; and the total volume of the solution is still 50 mL). After being mixed sufficiently, the quartz container was exposed 4 cm away under ultraviolet light until the reaction finished. All the reagents were analytical pure and the water was ultrapure.

2.2 Characterization

Images of the prepared nanoparticles were taken by transmission electron microscopy (TEM). The sample was prepared by dropping several drips of colloidal solution on carbon support film copper grid and dried at room temperature. Then the images were taken on H-800 TEM which operated at a voltage of 200 kV. By adopting lacey support film grid, HR-TEM detection was conducted, and the images were taken on Tecnai G^2 F20 field emission TEM operated at a voltage of 200 kV. Meanwhile, XPS was performed on the sample as well. The obtained nanoparticles were loaded on XC-72 carbon black for electro-chemical measurements based on the method reported in literature[7]. The electro-chemical measurement was conducted at 30℃, and three-electrode system was adopted with Pt as the counter electrode and SCE as reference electrode. The dissolved oxygen in the electrolyte was 0.5 $mol\cdot L^{-1}$ KOH + 2.0 $mol\cdot L^{-1}$ CH_3OH. Before the measurement, high pure N_2 was bubbled into the electrolyte to remove the dissolved oxygen and then CV curves were recorded in the N_2 atmosphere. The measuring conditions are as follows: potential is in the range -0.6~0.8 V, potential scan rate is 50 $mV\cdot s^{-1}$, and I-t curves are tested at electric potential of -0.3 V.

3 Results and discussion

3.1 Characterization of Pd@Pt nanoparticles

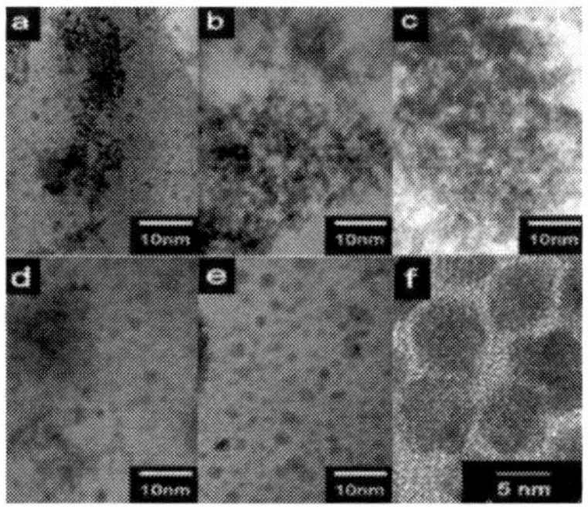

Figure 1. TEM and HR-TEM images of Pd core@Pt shell particles: Pd:Pt =4:1 (a), 1:1 (b), 1:4(c) by co-reduction; Pd:Pt =4:1 (d), 1:1 (e), 1:4(f) by Pd seed growth.

The TEM/HR-TEM and XPS were adopted to characterize the Pd@Pt nanoparticles synthesized by the two methods. Figure 1 shows the TEM of Pd@Pt nanoparticles synthesized based on photochemistry. In the figure, most particles are spherical nanoparticles with uniform size and good dispersion. All the average diameters of bimetallic nanoparticles synthesized by seed growth method are larger than the diameter of Pd seed (4.9 nm) and increase from 5.3 nm to 7.1 nm with the increase of Pt proportion; while the average diameters of the bimetallic nanoparticles synthesized by co-reduction reduces as the addition of Pt increases, decreasing from 4.1 nm to 2.9 nm. Figure 1f demonstrates the HR-TEM image of Pd@Pt core-shell composite particles which were prepared using seed growth method when Pd:Pt=1:4. Lattice fringes are observed clearly in the spherical Pd@Pt composite particles with 7 nm in diameter, which indicates that the composite particles are nano-crystalline.

Table 1. Results of the XPS analysis of Pd@Pt particles.

Pd:Pt (molar ratio)		4:1	1:1	1:4	4:1	1:1	1:4
Synthetic approach		Co-reduction			Seed growth		
Atom ratio	Pd %	70.6	40.0	2.1	70.4	40.2	3.4
on the surface	Pt %	29.4	60.0	97.9	29.6	59.8	96.6

XPS as an important surface analysis technology shows the surface atomic concentration and the chemical state of the Pd@Pt nanoparticles. It provides strong evidence for determining the structure of the particles. So, XPS was performed for the Pd@Pt nanoparticles synthesized using the two methods (Figure 2). The results are listed in Table 1. It indicates that the Pd@Pt nanoparticles synthesized using co-reduction and seed growth method shows similar XPS analysis and the molar ratio of Pd to Pt on the particle surface is much less than that of Pd (0) or Pd (II) to Pt (IV) in the solution during preparation. It can be speculated that there is higher Pt atomic fraction in the surface structure, which indicates that the prepared Pd@Pt particles have core-shell composite structure. When Pd:Pt=1:4, Pd element is hardly detected in the Pd@Pt core-shell nanoparticles synthesized by the both methods. This reveals that as the proportion of Pt increases, the Pt shell thickens continuously; in the condition, the Pt shell is too thick to make the inner Pd element be measured by XPS. In the linear fitting of the spectral lines of the samples in XPS energy spectrogram, two spectral peaks were observed in the Pt 4f and Pd 3d electronic spectrogram. Compared with the energy spectrum of standard palladium[8], the peak near 334 eV is simple substance Pd (Pd $3d_{5/2}$, 335.1 eV), and that at 336 eV around is oxidized Pd (PdO $3d_{5/2}$, 336.3 eV). It demonstrates that the Pt and Pd atoms on the surface of the synthesized Pd@Pt core-shell composite particles are oxidized to a certain extent.

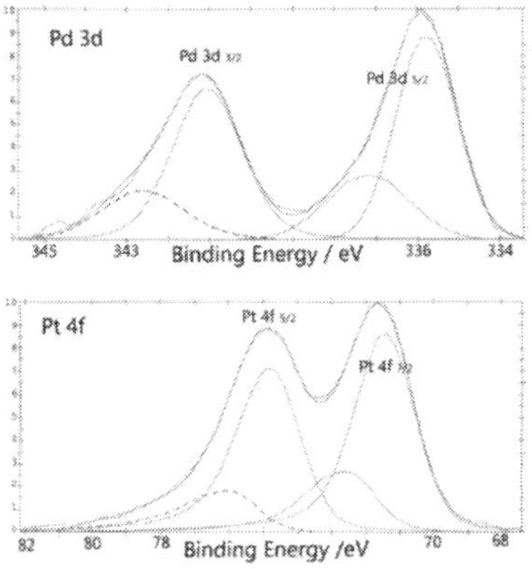

Figure 2. XPS spectra of Pd 3d and Pt 4f of Pd core@Pt shell particles by co-reduction (Pd:Pt = 1:1).

3.2 MOR activity of Pd@Pt core-shell nanoparticles

Figure 3 shows the cyclic voltammograms (CV) curves and time-current (I-t) curves of prepared Pd@Pt/C and commercial catalyst Pt/C (Johnson Matthey Company) for MOR in alkaline solution. Accordingly, the data of peak current density are listed in Table 2. The Pd@Pt core-shell nanoparticles synthesized using the two methods show high activity for MOR and similar activity law, that is, the activity at Pd:Pt=1:1 is higher than that at Pd:Pt=4:1 and 1:4. Additionally, the catalytic activity of Pd@Pt nanoparticles synthesized using co-reduction in Pd:Pt=1:1 and 4:1 conditions is higher than that synthesized using seed growth method in the same conditions. The highest catalytic activity was achieved for Pd@Pt nanoparticles synthezied by co-reduction when Pd:Pt=1:1. In the situation, the activity for MOR is 3 times of that of Pt/C catalyst, while the cost is reduced by about 1/4.

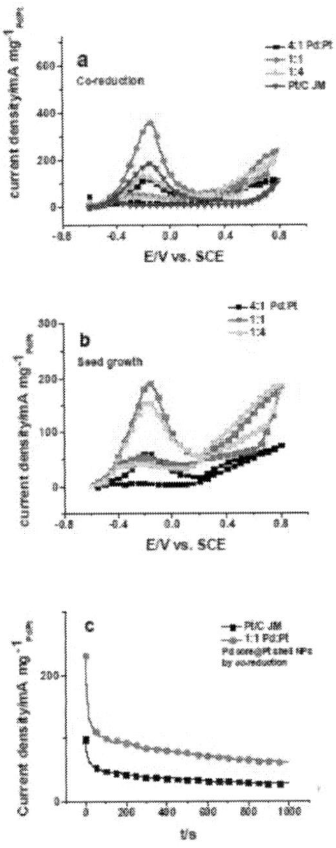

Figure 3. CV (a), (b) and I-t (c) of Pd@Pt/C in 0.5 mol•L^{-1} KOH + 2.0 mol•L^{-1} CH_3OH solution

Table 2. Results of the electrochemical tests of Pd@Pt nanoparticles

Synthetic approach		Co-reduction			Seed growth		
Pd:Pt (molar ratio)	Pt/C (JM)	4:1	1:1	1:4	4:1	1:1	1:4
Peak current density of CV (Am·mg$^{-1}_{PdPt}$)	161	115	360	135	61	160	157

According to the experiments, it is speculated that the Pd core promotes the catalytic activity of the synthesized Pd@Pt core-shell nanoparticles and the promotion is associated with the molar ratio of Pd to Pt. It is learned from the XPS analysis that the Pd in the Pd@Pt nanoparticles is prone to be oxidized. The oxidation products of Pd (PdO and PdOx) possibly promote the peeling of CO-like products which are generated during the reduction, therefore improving the catalytic activity of the composite nanoparticles[5]. On the other hand, the unique core-shell structure of Pd@Pt nanoparticles, which efficiently improves the utilization rate of Pt, is likely the other reason for the catalytic activity improvement of Pd@Pt/C core-shell catalyst. Owing to the smaller size and higher specific surface area of Pt, the nanparticles prepared based on co-reduction method show better electrocatalytic activity than that synthesized by seed growth method. As to the underlying reason for the high activity of Pd@Pt core-shell composite nanoparticles, it is possibly related to the fine structure of the nanoparticles.

Figures 3c illustrate the I-t curves of Pd@Pt/C. In continuous working condition, Pd@Pt/C core-shell catalyst (Pd:Pt=1:1, Co-reduction) exhibits similar stability to that of Pt/C.

4 Conclusions

(1) In PEG-acetone solution system, Pd@Pt core-shell composite nanoparticles with uniform size and good monodispersion were synthesized by photochemical reducing the mixture of Pd (II) and Pt (IV) ions or based on the Pd seed growth method with existence of Pt (IV). The size of the composite particles and the thickness of the Pt shell can be adjusted by changing the molar ratio of Pd (II) or Pd (0) to Pt (IV) ions. The average diameters of the obtained particles are in the ranges of 4.1~2.9 nm and 5.3~7.1 nm based on co-reduction and seed growth method, respectively. (2) XPS analysis reveals that the Pd@Pt nanoparticles synthesized by the two methods show core-shell structure. (3) Electrochemical analysis proved that all the synthesized Pd@Pt nanoparticles present high activity for MOR. The promotion of Pd core for Pt shell and the specific surface area of Pt increased by the core-shell structure are probably the main reasons for the high activity of Pd@Pt core-shell nanoparticles.

Acknowledgements

The work reported in this paper has received financial support from National Natural Science Foundation of China (No. 51164017 and 20863003).

References

1. X. D. Wang, X. F. Xie, M. Wang. Progress in Chemistry, **509**, 23 (2011)
2. J. Liu, S. Z. Qiao, J. Song. ChemComm, **13658**, (2011)
3. Y. N. Dong, S. S. Li, L. Cong. Yunnan Chemical Technology, **21**, 39 (2012)
4. Y. H. Cho, B. Choi. Electrochem. Commun., **378**, 9 (2007)
5. H. Wang, C. Xu, F. Cheng. Electrochem. Commun., **1575**, 10 (2008)
6. K. Sasaki, J. X. Wang, H. Naohara. Electrochem. Acta., **2645**, 55 (2010)
7. J. H. Zeng, J. Yang, J. Y. Lee. J. Phys. Chem. B, **24606**, 110 (2006)
8. F. M. John. *Handbook of X-Ray Photoelectron Spectroscopy* (Perkin-Elmer Physical Electronics DiVision, 1992)

MATEC Web of Conferences **65**, 03001 (2016) DOI: 10.1051/matecconf/20166503001
ICNSCM 2016

The Effects of Milling Factors on Surface Finish of Chromium-Molybdenum Steel

Warachanin Wongsasiripat[1] and Komson Jirapattarasilpls[1,a]
[1]*Department of Production Technology Education, Faculty of Industrial Education and Technology, King Mongkut's University of Technology Thonburi, Bangkok, Thailand*

Abstract. This research was to study effect of milling, which was affected on surface finish of Chromium-Molybdenum steel. Fifty-four pieces of JIS SCM 415 steel was used specimen. The experiment was done by CNC vertical milling machine and cutting tool was PVD coated tungsten carbide tool. The factorial design was applied and factors were consisted of three factors: cutting speed and feed rate and depth of cut that were set in three levels. The results revealed that surface finish was affected by two main factors of feed rate and depth of cut. On the other hand, cutting speed was not significantly main factor. However, interaction between cutting speed, feed rate and depth of cut were factors controlling surface finish. Finally, the effects of milling factors on surface finish were analyzed and discussed.

1 Introduction

Chromium-Molybdenum steel is alloy steel that contains low carbon. This steel uses for auto part and some machinery such as gear, shaft and piston. Because of chromium composition between 1% and Molybdenum about 0.15 % that steel have toughness and more tensile stress than other low carbon steel. To manufacturing of product, always this material was machined for milling process. This milling process was quality of dimension and surface finish that meet the needs of customer. Thus, factors of milling process could be affected to quality of machined surface. In previous study, factors of milling can be affected to quality of surface such as feed rate, cutting speed, depth of cut, cutting tool, cooling. The study of surface finish and effects of milling factors on surface finish have been report. The previous studies for surface finish of steel such as stainless steel [1], medium carbon steel [2-3] were studied. The others material has been studied such as titanium alloy [4-5], nickel alloy[5-6], aluminum alloy [7] and polymer [8]. They conclude that conditions of milling parameters based on feed rate, cutting speed, and depth of cut could be affecting to surface finish especially in case of steel. Therefore, Chromium-Molybdenum steel especially JIS SCM 415 steel was interested and effects of milling factors on surface finish for Chromium Molybdenum steel should be study.

2 Material, Tool and Equipment

Material used for this experiment was carbon steel that was grade JIS SCM 415. The properties of specimen were showed in Table 1. The specimens were prepared by dimension of 25 mm x 50 mm x 100 mm. as shown in Figure 1. The experiment was done by computerize numerically controlled milling machine: '*MIKRON model WF 21 D*'. The cutting tool was tungsten carbide that coated by physical vapor deposition (PVD) as shown in Figure 2. The cutting tool was installed into tool holder that diameter was 16 mm as shown in Figure 3. The surface roughness tester '*Mitutoyo Model SJ-201*' was used to measure surface finish.

Table 1. The Chemical Composition of JIS S50 C

Element	% wt.
C	0.13-0.18
Si	0.15-0.35
Mn	0.60-0.85
Cr	0.9-1.2
Mo	0.15-0.30

Figure 1. Specimen :JIS: SCM 415.

[a] Corresponding author: komson.jir@kmutt.ac.th

DOI: 10.1051/matecconf/20166503001

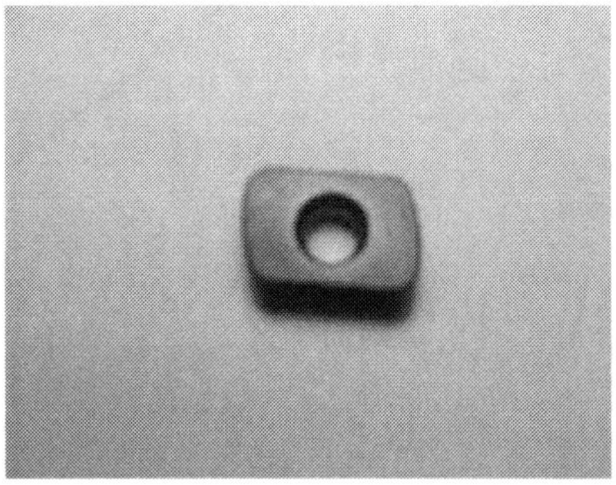

Figure 2. PVD coated tungsten carbide tool Specimen

Figure 3. Tool Holder.

3 Experimental Procedure

Experiment was applied by factorial design with experimental model of 3^3. The actual experiment was randomly selected run by two replicates which 54 trials. Milling factors used for this experiment were cutting speed (v), feed rate (f) and depth of cut (d). Each factor was set at three levels. Cutting speeds (V) were 50, 100, and 150 meter per minute. Feed rate was set at 100, 200, and 300 millimeter per minute. Depth of Cut was 0.2, 0.4 and 0.6 millimeter. These factors were tested by pilot study before running the actual experiment.

The surface finishes of each specimen were measured by surface roughness average (Ra). The measuring was done by the three times on machined surface. Data were analyzed by statistical methods to finding the factors affected to surface roughness.

4 Results and Discussion

The results of surface finish were observed and measured were showed as example in Figure 4. The initial setting factors to run for appropriate levels of milling were determined and surface roughness results were tested by the normality test. The test showed that the results were normality.

Effect of milling factors on surface finish was anylzed by the analysis of variance (ANOVA). ANOVA indicated that the effect of feed rate and depth of cut were significantly difference at the confidant level of .01. It means that feed rate and depth of cut were main factors that directly affected to surface finish. Furthermore, cutting speed was not main factor. However, interaction between each factor was found. That means interaction between three factors would be affect to surface roughness.

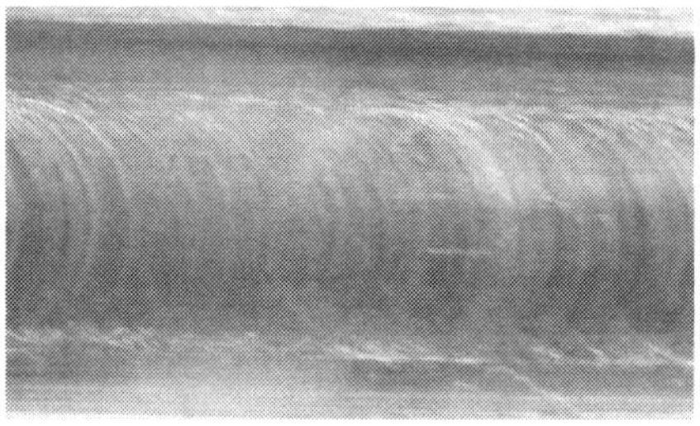

Figure 4. Surface Finish of Specimen
Condition : V= 50 m/min, f = 300 mm/min d = 0.20 mm
Ra = 1.861

4.1 The effects of feed rate and depth of cut on surface finish

Feed rate and depth of cut were main factor affected to surface roughness, as shown ANOVA. The effect of feed rate was revealed that feed rate was higher, surface roughness was slightly up as shown in Figure 5. It means that the low feed rate was affected to good quality of surface. Meanwhile, the effect of depth of cut was shown in Figure 6. The increasing depth of cut was affected to enlargement of surface roughness. That means milling at higher depth of cut, it will be made poor quality of surface finish.

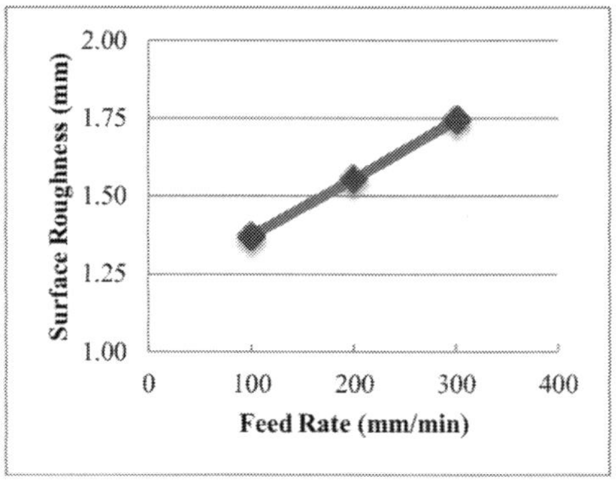

Figure 5. Effect of Feed Rate.

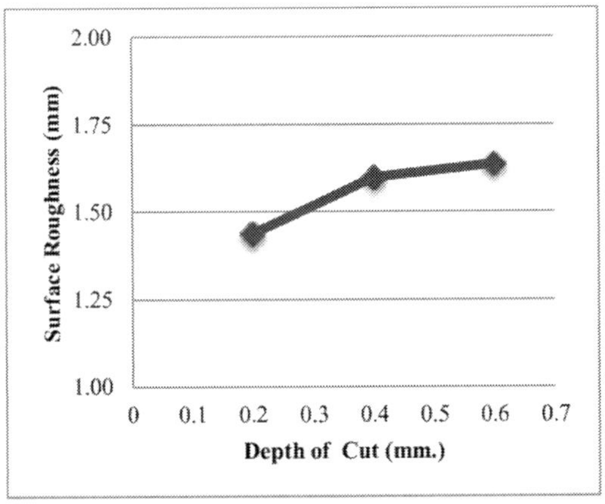

Figure 6. Effect of Depth of cut.

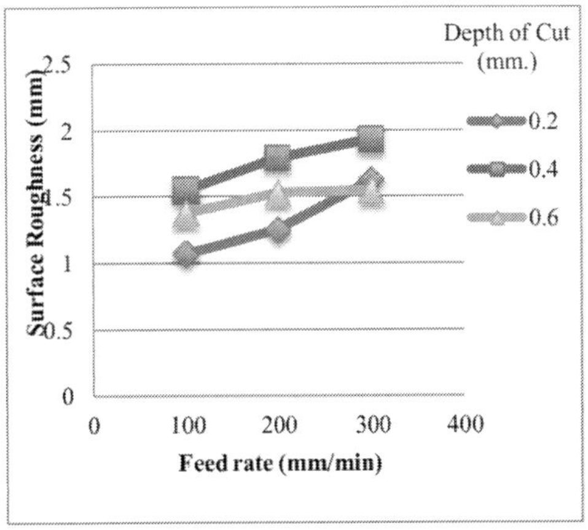

Figure 8. Effects of interaction of milling factors at Cutting speed 100 m/min

4.2 The effects of interaction of milling factors on surface surface

Since the cutting speed was low that the surface roughness was lowest when applied feed rate and depth of cut at low level as shown in Figure 6. As same as low cutting speed, medium cutting speed showed that using feed rate and depth of cut at low level could be produce best surface finish as showed in Figure 7. On other hand, at high cutting speed, surface finish was lowest at high depth of cut and low feed rate as showed in Figure 8. In order to good quality of surface finish, it was recommended to use low feed rate and use low cutting speed with low depth of cut but if it needs to use high cutting speed then use high depth of cut too.

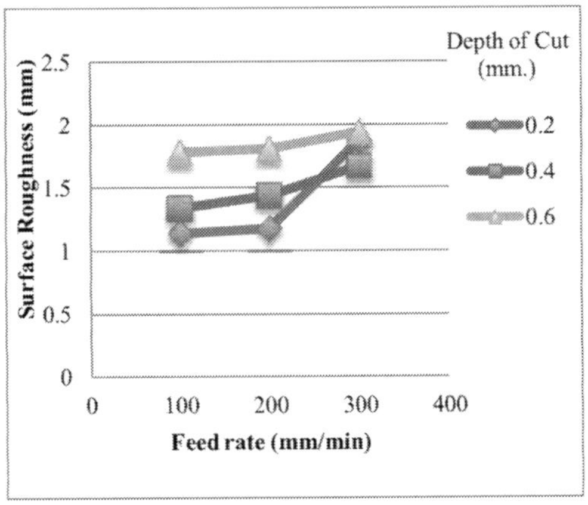

Figure 7. Effects of interaction of milling factors at Cutting speed 50 m/min

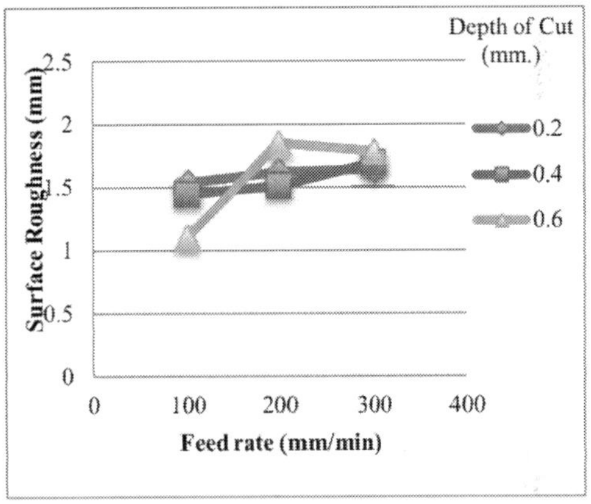

Figure 9. Effects of interaction of milling factors at Cutting speed 150 m/min

4 Conclusion

This research was experiment of effects of milling parameters on surface finish of Chromium-Molybdenum steel, JIS SCM 415, that was milling by coated tungsten carbide cutting tool. The results could be conclude that feed rate and depth of cut were main factors that affected to surface finish. Moreover, interaction of factors was found that interaction effects represent the combined effects of each factor on surface finish. The impact of cutting speed depends on the level of the feed rate and depth of cut. Moreover, the research was concluding as following;

- The low feed rate was affected to good surface finish better than high feed rate.
- The lower depth of cut will be made a good quality of surface finish more than higher.

- The low cutting speed with low depth of cut and low feed rate can be produce good surface finish as same as cutting speed at high level with high depth of cut and low feed rate.

References

1. S.Luejanda and K.Jirapattarasilp, *The Study of Surface Finish in Face Milling of Stainless Steel:AISI 304*, Adv. Mater. Research, Vol. 650 (2013), p.606.
2. P.Chutrakul, K.Jirapattarasilp, *The Effect of Hard Milling on Surface Finish of Medium Carbon Steel,* Adv. Mater. Research, Vol. 904 (2014), p.213.
3. R.J.Wang, Y. L.Zhang, H.B.Liu, M.H.Du, *Surface Finish Study of Hard Milling 45 Steel with PCBN Tool*Adv. Mater. Research, Vol. 468-471 (2012), p.1467.
4. K.Tsuda, K.Okuda, H. Shizaka, and M. Nunobiki, *A Study of the Micro End Milling of Titanium Alloy*, Adv. Mater. Research, Vol 325(2011), p.588.
5. D.Ulutan,T. Ozel, *Machining induced surface integrity in titanium and nickel alloys: A review* International Journal of Machine Tools and Manufacture, Vol.51 (3) (2011) , p. 250.
6. M.A.Hadi, J.A.Ghani, C.A.Haron, N.A. Raof, *Tribological Characterization of the Cutting Zone in Milling Nickel Based Alloy,* Appl. Mech. and Mater. Vols. 554 (2014) p.12
7. J.Abbas, A.A.Habaibeh,and D. Z.Su, *The Investigation of Prediction Surface Roughness from Machining Forces in End Milling Processes*, Key Eng. Mater. ,Vol.486, (2011), p 91.
8. S.Obma and K.Jirapattarasilp, *The Effect of Vertical Milling on Finished Surface of Polymethyl-methacrylate Sheet,* Appl. Mech. and Mater. Vols. 157-158 (2012) p.263

Temperature Control of Heating Zone for Drying Process: Effect of Air Velocity Change

Chananchai Wutthithanyawat [1,a] and Nawadee Srisiriwat [2]

[1]*Department of Instrumentation and Control Engineering, Pathumwan Institute of Technology, Bangkok 10330, Thailand*
[2]*Department of Petrochemicals and Environmental Management, Pathumwan Institute of Technology, Bangkok 10330, Thailand*

Abstract. This paper proposes a temperature control technique to adjust air temperature in a heating zone for drying process. The controller design is achieved by using an internal model control (IMC) approach. When the IMC controller parameters were designed by calculating from an actual process transfer function estimated through an open-loop step response with input step change from 50% to 60% at a reference condition at air velocity of 1.20 m/s, the performance of temperature controller was experimentally tested by varying an air velocity between 1.32 m/s and 1.57 m/s, respectively. The experimental results showed that IMC controller had a high competency for controlling the drying temperature.

1 Introduction

Drying process is widely used for food processing to preserve foods by removing enough moisture from foods to prevent food spoilage in which effective drying depends on: adequate heat to draw out moisture, without cooking the food; dry air to absorb the released moisture; and enough air circulation to remove the moisture [1]. Air flow rate and temperature in drying chamber affect the moisture removal rate from foods [2]. Consequently, the air flow rate and temperature control is very important to generate hot air in heating zone of food drying process.

In our previous work [3], the temperature distribution of hot air flow in heating zone of a rectangular duct was investigated by varying the power supply to a coil heater and an electrical fan in order to generate hot air entering the drying chamber that is crucial to control the desired temperature and keep the quality of air for drying process. This experiment installation associated with a data acquisition (DAQ) card and a computer was also conducted to generate user interfaces on the computer by using LabVIEW [4].

Currently, there are many control techniques used for the food drying system such as herb [5], gain [6], codfish [7-8], and fruit [9]. Nafeh et al. [5] investigated a temperature control in a PV-Wind medical herb dryer system by using on/off controller that showed the high capability of the proposed technique in controlling the drying temperature. Han et al. [6] used a model predictive control (MPC) system based on neural network which was designed and tested on two model corn dryers that the control method of the system can be used in drying process control. Boeri et al. experimentally investigated the optimized operating drying control for temperature, relative humidity and air velocity parameters by using fuzzy logic approach [7] and also proposed the multi-input-multi-output-based PID and Fuzzy logic (FLC) controller for a drying control process [8]. Geddes [9] implemented an internal model control (IMC) model to the temperature controller for fruit dryer process in which the simulation results showed that IMC controlled system could achieve the requirement of providing offset-free responses at steady state. Although there are many works concerning a controller design for the food drying process, the research relating to the experimental installation and controller design using IMC for the drying system is limited.

In this paper, the drying temperature controller was designed by using IMC to adjust the desired temperature by manipulating the power supply of coil heater used for heating air in the heating zone of drying process. Because the foods are sensitive to heat, there is a suitable drying temperature for each food and the air flow rate is another important parameter varied to remove moisture to be the required value at the fixed temperature. Consequently, the temperature response with varying air velocity was investigated and the capability for temperature controlling was experimentally tested.

2 Experimental Installation

In the heating zone, the main apparatuses are composed of a 500 W coil heater, supplied with the input signal in the range of 0 to +5 VDC to generate the maximum temperature about 102.8 °C, and an electrical fan (EBM-PAPST, 8414N/2GH), adjusted the input signal in the range of 0 to +5 VDC to provide variable air velocity between 1.20 and 1.57 m/s corresponding to the air

[a] Corresponding author: chananchai.wut@pit.ac.th

flowrate from 0.87 to 1.14 m^3/min. The atmospheric air forced by electrical fan flows across the coil heater to generate hot air passing through the non-insulated rectangular duct as shown in Fig. 1. The air velocity was measured by using a Panasonic SUNX Photo Sensor PM-K44. The temperatures inside the rectangular duct at various positions were measured by RTD (resistance temperature detector). The bottom of apparatus is the circuits of measurement and control system and the output terminal can be used to send the signal and then monitor the temperature of three positions in rectangular duct and the air velocity via LabVIEW [4].

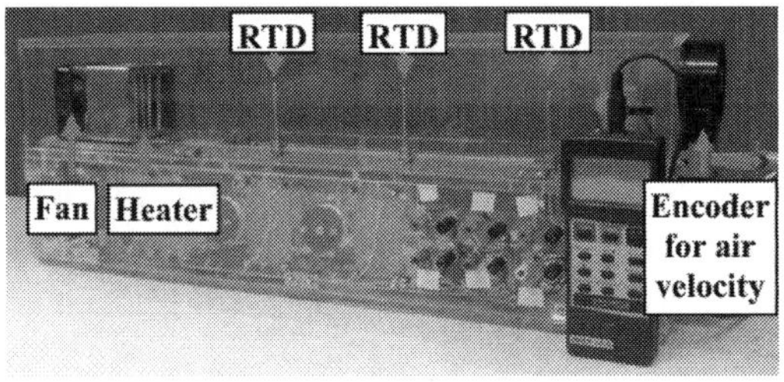

Figure 1. Experimental apparatus [3].

3 Design of Temperature Control

It is known from the previous discussion that the main function of the electrical heater is to supply the heating required by the drying air. The block diagram of thermal process based on two-input single-output (TISO) system is illustrated in Fig. 2 which includes the temperature control system and the air velocity control system.

When the open-loop system was provided with a step input, the process step testing was carried out to characterize three parameters, the gain K, the dead time t_0 and the time constants τ, for determining the first-order plus dead-time (FOPDT) model. Accordingly, the combination of the transfer functions, $G_P(s)$, is approximated by FOPDT model for the purpose of characterizing the dynamic response of the process. In this step testing, to obtain the maximum gain value, the change of 10% of step input from 0 to 90% was tested for RTD#1 and it was found that at a step change, started at 100 sec, from 50% to 60% gave the highest value of steady-state gain in which the process variable was changed from 49.62% to 65.13% or from 62.21°C to 73.85°C for the step change in controller output of magnitude is 0.5 V. Therefore, the steady-state gain of process, K, can be calculated as 23.27°C/V corresponding to 1.47%/%. Using Fit 1 method to determine the dead time and time constant [10], the values of t_0 and τ are approximately 15 sec and 150 sec, respectively. Thus, the resulting FOPDT model transfer function of actual process, $G_P(s)$, is shown in Eq. (1).

$$G_P(s) = \frac{Ke^{-t_0 s}}{\tau s+1} = \frac{1.47e^{-15s}}{150s+1} \quad (1)$$

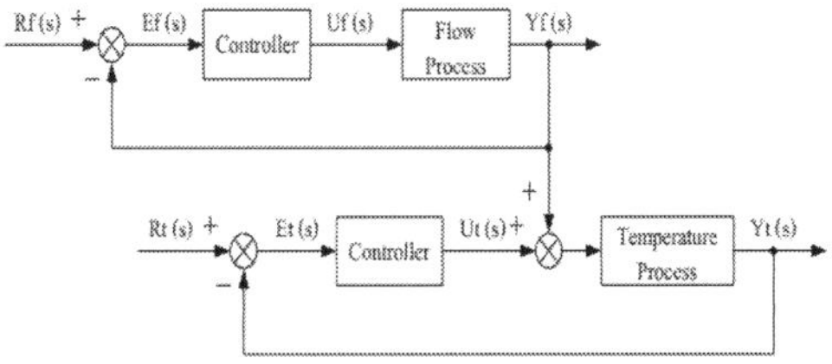

Figure 2. TISO Block diagram of the temperature control system coordinated air flow control system.

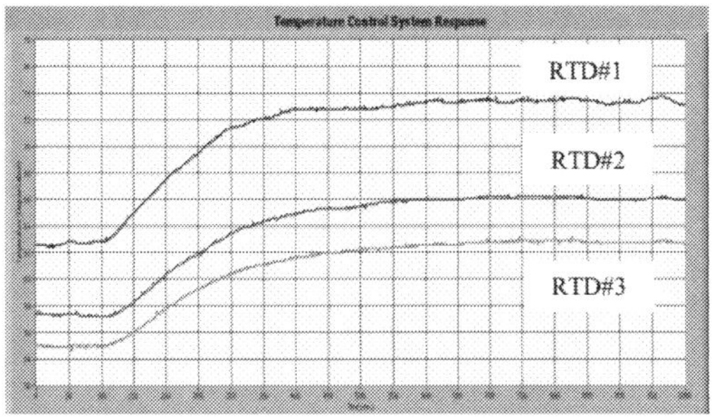

Figure 3. Open-loop step response of temperature process from 50% to 60%.

In process control applications, model based control systems are often used to track set points and reject low disturbances. Particularly, if the control structure has been developed based on the exact model of the process then perfect control is theoretically possible. For controller design in this paper, IMC based PI (proportional integral) design procedure for a first order system was applied to control air temperature at the RTD#1 position and the fixed air velocity of 1.20 m/s. The PI tuning parameters of controller were calculated by Eq. (2) [11]. Hence, the tuning parameters of K_c and T_i are 1.13 and 150 sec, respectively.

$$K_C = \frac{T}{K(T_f + t_0)} \; ; T_i = T \quad (2)$$

where K_c is controller gain [%/%], K process gain [%/%], T_i integral time constant [sec], T_f filter time constant [sec], t_0 dead time [sec] and T is process time constant [sec]. In this paper, as a rule of thumb T_f is 75 sec chosen to be at least twice as fast as the time constant from open-loop response of process [9].

4 Results and Discussion

The aim of the temperature control system is to adjust the temperature in heating zone of drying chamber at which the range of temperature and air velocity varying in this study was generally used in the drying foods. After the IMC controller parameters were designed, to investigate the performance of the proposed temperature control system, the response of the drying temperature was tested at different air velocities between 1.32 and 1.57 m/s corresponding to the step change of input signal from 20 to 100% of +5 VDC for supplying the electric fan.

Figs. 4(a) and 4(b) illustrate the temperature response and the control signal, respectively, in case of using IMC controller. In this test, a set point of drying temperature was nearly 62.52°C which is at 50% operating point of temperature control system. The change of air velocity entering the heating zone in drying chamber was assumed as a disturbance of drying process. As shown in Fig. 4, the step changes of input signal of the electric fan at 20, 40, 60, 80 and 100%, representing the air velocity at 1.32, 1.41, 1.49, 1.55 and 1.57 m/s, respectively, were started at 100 sec. An ambient temperature was about 25°C.

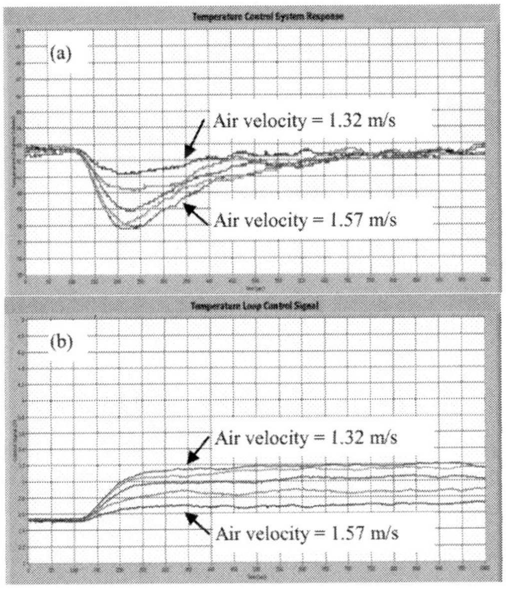

Figure 4. Effects of air velocity on (a) temperature control system response; x-axis is time [sec] and y-axis is temperature [°C] and (b) temperature loop control signal; x-axis is time [sec] and y-axis is control signal [V].

At step change of 20%, when the air velocity was changed from 1.20 m/s to 1.32 m/s, the minimum temperature was about 61.14°C (about 48.0%) which decreases to 2.0% from the set point and the control system took a settling time of 350 sec to come within the prescribed value and remain within the acceptably value as shown in Fig. 4(a). The control signal was about 54% which is more than 50% of a reference operating condition, that is, the control signal increased 4% as illustrated in Fig. 4(b). If the inlet air velocity increases, thus making a disturbance, its effect must propagate through the entire heating zone before the drying temperature changes. Once the required value of drying temperature changes, the signal from the controller also changes. An increase of air velocity decreased the minimum temperature response. At 100% step change of air velocity, the minimum temperature was at 57.85°C or about 43.8% which decreased to 6.2% from set point that the control signal was about 64%. The settling time increased with increasing the air velocity to reach the set point of drying temperature and moreover, the increase of air velocity in range of 60 to 100% used the highest settling time about 600 sec. Although the temperature control system took a long time to control the desired drying temperature because this is the worst case of a non-insulated system in order to test an ability of control system, the drying temperature was efficiently controlled nearly set point. Consequently, it can be concluded that the control of drying temperature was achieved through using IMC controller which gave a high capability.

5 Conclusion

The temperature control system of heating zone in drying chamber was performed by using internal control model (IMC). The drying temperature and air velocity in range of testing in this paper are usually applied in food drying process. The IMC based PI control was designed to adjust the drying temperature at the required value at approximately 62.52°C (50% operating condition) at 1.20 m/s of air velocity to control the operation of the electric heater. To investigate an efficiency of control system, the air velocity, assumed as a disturbance, was varied between 1.32 and 1.57 m/s. The experimental results indicated that the proposed IMC based PI controller gave a good capability in controlling the drying temperature.

References

1. P. Kendall, P. DiPersio, J. Sofos, Food and Nutrition Series/Preparation **9.308**, (2012)
2. M.A. Tabassum, V.K. Jindal, Pakistan Journal of Agricultural Research **13**, 312-319 (1992)
3. C. Wutthithanyawat, N. Srisiriwat, Applied Mechanics and Materials **627**, 153-157 (2014)
4. C. Wutthithanyawat, N. Srisiriwat, Applied Mechanics and Materials **619**, 303-308 (2014)
5. A. Shafy, A. Nafeh, E.A. Sweelem, F.H. Fahmy, Smart Grid and Renewable Energy **4**, 153-160 (2013)
6. F. Han, C. Zuo, W. Wu, J. Li, Z. Liu, Mathematical Problems in Engineering Article ID 584376, (2012)
7. C. Boeri, F. Neto da Silva, J. Ferreira, M. Soares dos Santos, *European Drying Conference - EuroDrying'2011, Spain*, (2011)
8. C.N. Boeri, F.N. da Silva, J. Ferreira, Acta Scientiarum. Technology **35**, 279-289 (2013)
9. S.A. Geddes, *Internal Model Control (IMC) of a Fruit Drying System, Bachelor of Engineering dissertation, Faculty of Engineering and Surveying, University of Southern Queensland*, (2006)
10. C.A. Smith, A.B. Corripio, *Principles and Practice of Automatic Process Control*, John Wiley & Sons, Inc., Canada, (1985)
11. R. Singh, R. Bala, B. Bhatia, International Journal of Advanced Research in Computer Science and Software Engineering **4**, 915-922 (2014)

Numerical Investigation on Mechanical Behaviour of Closed-cell Aluminium Foams Using a Representative Volume Element Method

Chengjun Liu[1], Y.X. Zhang[1], Chunhui Yang[2*]

[1]*School of Engineering and Information Technology, UNSW Canberra, Australian Defence Force Academy, ACT 2600, Australia*

[2]*School of Computing, Engineering and Mathematics, Western Sydney University, Penrith, NSW 2751, Australia*

Abstract. In this study, a representative volume element (RVE) method is developed to capture the mechanical behaviour of aluminum foams under compressive loadings. Octadecahedrons are selected for forming cells of a microstructured RVE model to simulate mechanical behaviour of the aluminium foams under quasi-static compressive loadings. A convergence test is conducted to determine an appropriate mesh density. The stress-strain relationship obtained from the numerical simulations is compared to that from experimental study and agreements between these results demonstrate the efficiency of the proposed RVE model. Thereafter failure modes during a compressive loading process of the aluminum foams are also identified and discussed using this RVE model.

1 Introduction

Closed-cell aluminium foams are very competitive materials for applications in engineering structures where weight and impact resistance are the main concerns, due to their superior mechanical characteristics such as light weight and energy absorption. Typical engineering applications of aluminium foams include automotive vehicles, aeronautical structure, especially on helicopter components, aerospace structures, ship building, railway train and packaging components [1-3]. To facilitate the structural applications of the aluminium foams with an effective and optimal structural design, it is essential to gain an insightful understanding of their mechanical behaviour.

Multiscale modelling is a term for a kind of modelling in which multiple models at different length and/or time scales are developed and employed simultaneously to describe an engineering problem of a material or system. However it is difficult to achieve such a systematic algorithm across both length and time scales. To simplify this complex problem with ignoring the influence from time scale, a systematic framework on multiscale analysis and digital material representation of heterogeneous material [4], which are multiphase and microstructured, has been developed and ustilised to fabricate and evaluate their overall material properties, i.e., porous materials and particle-reinforced composites. It is based on micromechanics and theory of localisation and homogenisation. This digital framework can be employed to cover three-level digital material representations (DMR) across length scales from micro, meso to macro scales: a) unit cell (UC); b) representative volume element (RVE); and c) bulk material. In Yang et al.'s study [4], the representative volume element (RVE) models for metal foams were generated using a microstructure generator coded using Matlab with a sphere growth algorithm. Then the finite element modelling was employed to determine their mechanical properties. Other finite element models [5-6] have been also developed and employed to model the metal foams and study their mechanical behaviour, but these models were mainly generated based on a well-known Voronoi tessellation method, which may result in under- or over-estimation on mechanical properties due to the randomness and need an averaging post-processing.

Utilising repeating polyhedron cells is a possible way to overcome this flaw and can provide a reasonable estimation without one more post-processing step. Researchers have developed a few finite element models based on tetrakaidecahedron cells for modelling of aluminum foams. Simone and Gibson [7] proposed a model and analysed the effect on stiffness and strength of closed-cell tetrakaidecahedral foams. Song et al. [8] developed a model with tetrakaidecahedron cells to model the dynamic crushing behaviour of aluminium foams. Nonetheless, numerically modelling the mechanical properties of aluminum foams using uniform and closed-cell polyhedrons is still rare.

In this research, a Representative Volume Element (RVE) method is proposed to model the closed-cell aluminium foams focusing on their mechanical behaviour under compressive loading. Octadecahedrons are selected as basis cells to form the RVE model for the aluminium foams. For validation and verification, this microstructured RVE model is employed to model Alporas® foams with a porosity of 0.915 under quasi-

*Corresponding author: r.yang@westernsydney.edu.au

static compressive loading with a loading rate of 2 mm/min.

2 RVE model of Closed-Cell Foams

This geometric model of the proposed RVE was obtained through cutting 8 octadecahedrons as shown in Figure 1, including one intact octadecahedron wrapped in the center, twelve half-octadecahedrons (four on the top, four in the bottom and another four surrounding the intact cell) and four quarter-octadecahedrons in parallel.

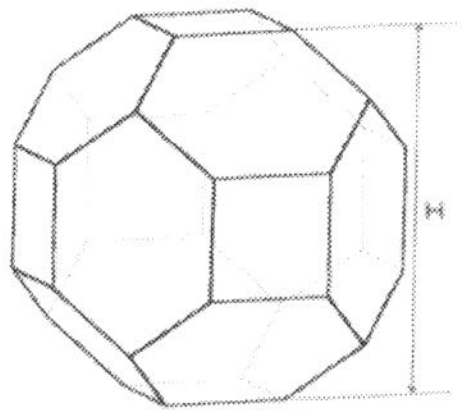

Figure 1. Octadecahedron.

As octadecahedral foams can be accounted as a number of repeated RVEs in the structure, this RVE is considered to represent aluminium foams with uniform and closed octadecahedron cells. The height is used to define the cell size of the octadecahedron and specifically the height H is fixed to 2.96 mm in this study while the thickness of the cell t is 0.034 mm to guarantee that the relative density of the foam is approximately 8.5%, which is equivalent to the porosity of 91.5% for the aluminium foam. The geometric model is shown in Figure 2 with the RVE defined as Part 1, and the upper and bottom steel plates defined as Part 2 and Part 3, respectively. Part 3 is defined to support the RVE, while compressive loads are applied to the RVE via moving Part 2.

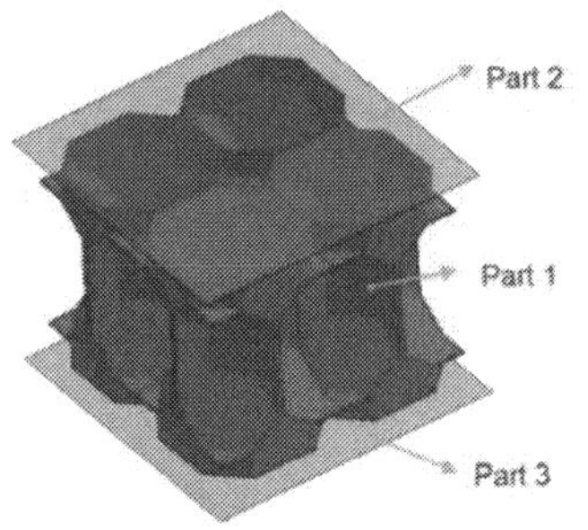

Figure 2. Geometric model of the RVE.

3 FEA of the RVE model

This finite element analysis was conducted using commercial finite element analysis software LS-Dyna. Parts 1 and 2 in the FE model were meshed using the Belytschko-Tsay shell elements, with setting a shear correction factor of 5/6, and the number of through shell thickness integration points to 1. The RVE of foams was meshed using 3D solid element 164, a fully integrated quadratic 8-node element with nodal rotations. It was divided into eight equal parts as shown in Figure 3a, and each part was divided into hexahedrons (see Figure 3b) and each hexahedron can be meshed using n × n solid elements.

The material model of aluminum foams used a low cost isotropic plasticity model (Material Type 12). In this model, the strain rate is considered using the Cowper-Symonds model, which scales the yield stress via a strain rate dependent factor [9]. The steel plates, Part 1 and Part 2, were both treated as rigid body in the FE model. Material parameters for these two metals are given in Tables 1 and 2, respectively. Two types of contact boundary conditions were defined in this model. One is the mutual contact between different parts, specifically between the RVE and Part 2 or Part 3. The other one is self-contact inside the RVE model resulted from large deformation.

Table 1. Material properties of the aluminium [2, 10].

Density	Shear modulus	Yield stress	Plastic hardening modulus	Bulk modulus
2700 kg/m^3	23.73 GPa	100 MPa	1.36 GPa	51.42 GPa

Table 2. Material properties of the steel [11].

Density	Young's Modulus	Poisson's ratio
7850 kg/m3	210 GPa	0.33

Convergence test was conducted to find out the best value of n as shown in Figure 3 (b). The compressive force-time curve was computed from the finite element model with different mesh density with n ranging from 1 to 5. According to Figure 4, converged results are achieved with the mesh density of n = 5. Therefore n = 5 was adopted to mesh the RVE model for accuracy and computation efficiency. The finite element mesh using this mesh density is shown in Figure 5.

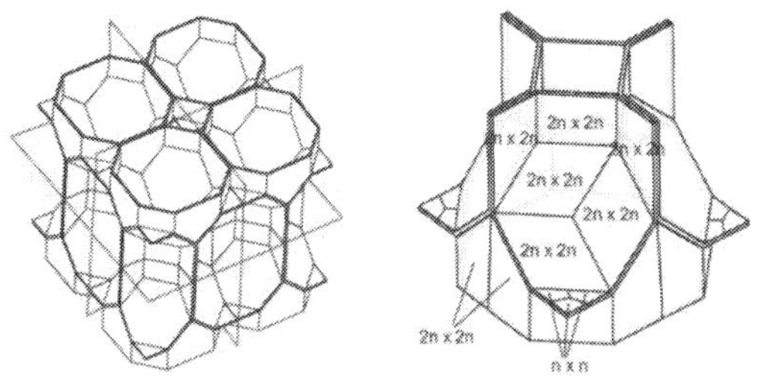

(a) (b)

Figure 3. Mesh refinement of the RVE model.

DOI: 10.1051/matecconf/20166503003

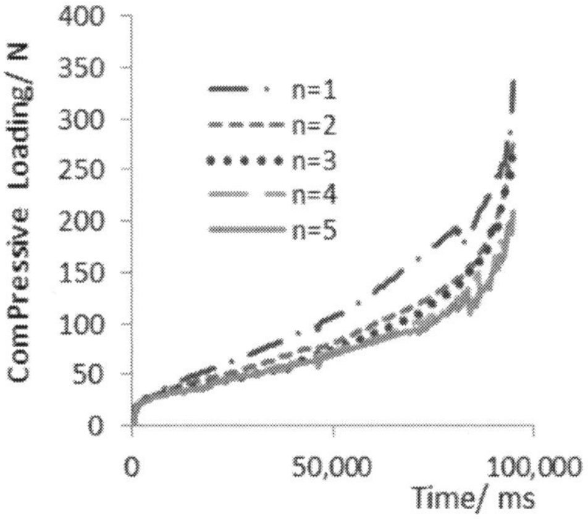

Figure 4. Mesh refinement of the RVE model.

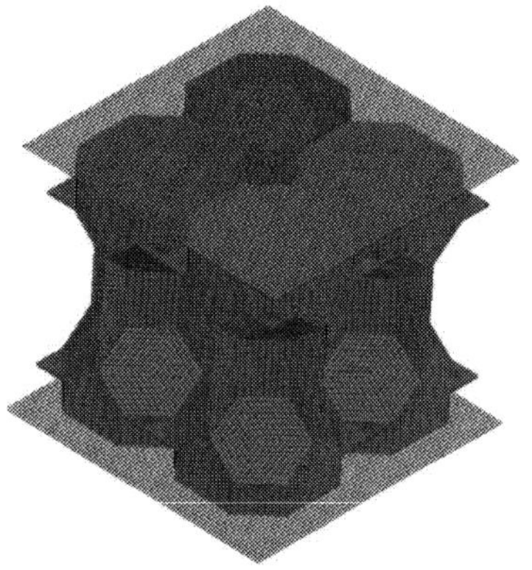

Figure 5. Finite element mesh of the RVE model.

4 Results and discussion

As for validation and verification, the stress-strain relations extracted from the RVE model were compared to that from a compression test of equal-porosity Alporas® foams.

Figure 6 compares the compressive stress-strain curves of the aluminium foams obtained from the finite element analysis and the experimental study, respectively [12]. The curve obtained from numerical modelling presents a trend very similar to that from the experiments and the model can capture the hardening behaviour of the material, though there is a discrepancy between the numerical result and experimental data. The RVE model provides lower Young's modulus and yielding stress than those from the experimental work, while the plateau stress in the modelling is higher than that from experimental data. These differences may result from the employing uniform cell and cell size to represent the aluminium foams. However, it shows a probability that the aluminium foams with a uniform cell could present better mechanical properties than aluminium foam with random pore sizes. Overall, the RVE-based finite element model can reflect the mechanical properties of closed-cell aluminium foams.

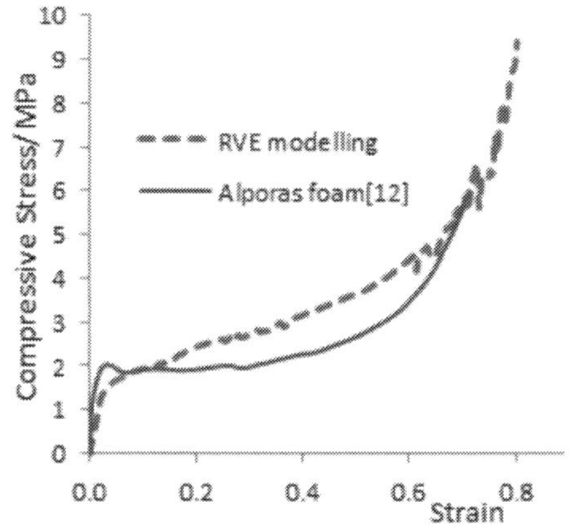

Figure 6. Stress-strain curves of the aluminium foams.

The modelled deformation evolution of the RVE during the compression process is shown in Figure 7. Several failure modes could be observed during the deformation process. The first failure mode is cell wall buckling, which can be clearly found from the cell walls along the compressive loading direction. The second type of the failure modes is stretching, which occurs in the cell walls of both top and bottom semi-cells, in which their normal directions to align with the loading direction. Another failure mode is upsetting in cell walls, where the vertical cell walls become wider and shorter when the compressive loading is applied.

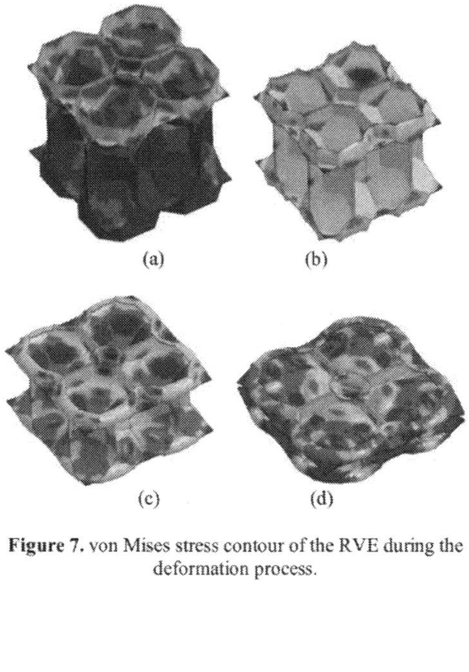

Figure 7. von Mises stress contour of the RVE during the deformation process.

5 Concluding Remarks

A RVE model based on repeated octadecahedrons has been developed in this study to model the mechanical behaviour of closed-cell aluminum foams numerically. A convergence test has been conducted and the results show that the finite element model converges well.

The compressive stress versus strain curve of the aluminum foams obtained using the proposed RVE model of the Alporas® aluminum foams has been compared to that obtained from experiment. Overall, the RVE-based finite element model can capture the compressive mechanical properties of closed-cell aluminium foam. The deformation process and the failure modes of the RVE during the compressive loading process from the numerical modelling are also presented, in which in total three failure modes have been identified and they are buckling, stretching and upsetting, respectively.

The proposed RVE model can be further extended to model other metal foams considering different porosities and different basis metal materials as part of future work. The strain rate effect could be investigated by using this model as well.

References

1. L.J. Gibson and M.F. Ashby: *Cellular Solids Structure and Properties* (Cambridge University Press, 1999).
2. M.F. Ashby, A.G. Evans, N.A. Fleck, L.J. Gibson, J.W. Hutchinson and H.N.G. Wadley: *Metal Foams: A Design Guide* (Butterworth-Heinemann, Boston 2000).
3. J. Banhart: Prog. Mater. Sci. Vol. 46 (2001) p. 559–632.
4. C. Yang, Y. An, M. Tort and P.D. Hodgson: Comput. Mater. Sci. Vol. 81(2014) p. 89–97.
5. K.R. Mangipudi and P.R. Onck: J. Mech. Phys. Solids Vol. 59 (2011) p. 1437-1461.
6. K.R. Mangipudi and P.R. Onck: Acta Mater. Vol. 59 (2011) p. 7356-7367.
7. A.E. Simone and L.J. Gibson: Acta Mater. Vol. 46 (1998) p. 2139-2150.
8. Y. Song, Z. Wang, L. Zhao and J. Luo: Mater. Des.Vol.31 (2010) p. 4281-4289.
9. John O. Hallquist: *LS-DYNA Theory manual* (March 2006).
10. I. Jeon, K. Katou, T. Sonoda,T. Asahina and K.J. Kang: Mech. Mater. Vol. 41 (2009) p. 60-73.
11. W.H. Hou, F. Zhu, G.X. Lu and D.N. Fang: Int. J. Impact Eng. Vol. 37 (2010) p. 1045-1055.
12. J.K. Hu, G.X. Lu and D. Ruan: Composites Part B Vol. 41 (2010) p. 678-685.

Inhibitory Effect of *Cratoxylum formosum* Gum on *Candida glabrata* and Its α-mangostin Content

Sroisiri Thaweboon[1,a], Boonyanit Thaweboon[1,b], Passiri Nisalak[1,c] and Rattiporn Kaypetch[1,a]

[1]*Faculty of Dentistry, Mahidol University, Thailand*

Abstract. *Candida glabrata* is the most common fungal species isolated in patients with severe mucosal inflammation. The high resistance to traditional antifungal therapies makes this species a growing concern in clinical settings. *Cratoxylum formosum* is a plant widely distributed in mountainous area of Asian countries. This study aims to examine antifungal activity of *C.formosum* gum against *C.glabrata* and its α-mangostin content. Inhibition of fungal growth was primarily tested by agar diffusion. Broth dilution method was then used to determine the minimum inhibitory concentration (MIC). The α-mangostin content was determined by high performance liquid chromatography (HPLC). Inhibitory effect of the gum was seen against *C.glabrata* (clinical isolate and ATCC22019) with zones of inhibition ranging from 14.3 to 10.2 mm. MIC value against *C. glabrata* ATCC22019 and the clinical isolate was 1.25 mg/mL. By HPLC, the α-mangostin content of *C.formosum* gum was determined as 4.08% (w/w). In conclusion, the anticandidal activity of *C.formosum* gum suggests that this plant may be a useful source for the development of a novel antifungal agent against candidal infection. Further *in vitro*/*in vivo* studies should be conducted to understand the mechanisms of action and to establish the safe profile of this gum for clinical usage..

1 Introduction

Candida glabrata is normally commensal of humans that can be found especially in the oral cavity and the gastrointestinal tract of most healthy humans. However, it can turn into opportunistic pathogen causing mucosal and blood stream infections in persons with predisposing factors, such as treatment with antibiotics, diabetes, cancer, extreme age, immunosuppression or long-term hospitalization [1]. Apart from being the most common non-*Candida albicans* (NCAC) species isolated in oral cavity of patients with severe mucosal inflammation, *C. glabrata* exhibits the high resistance to traditional antifungal therapies [2].

Cratoxylum formosum is a plant widely distributed in mountainous area of various Asian countries. The gum extracted from the burned bark has been used to stain on tooth surfaces to prevent tooth pain, tooth decay or other oral diseases [3]. Previous studies have shown the antimicrobial properties of *C. formosum* gum against tooth decay associated bacteria and *Candida albicans* [4,5]. However, the effect on *C. glabrata* does not exist. Therefore, the objective of this study was to examine the antifungal activity of *C. formosum* gum against *C. glabrata in vitro*.

2 Materials and Methods

2.1 Plant materials preparation

The trunks of *Cratoxylum formosum* were collected from mountainous regions of northern Thailand. They were cut and burned to obtain a black colored extract, which was immediately collected and used to prepare the gum.

2.2 Candida preparation

Candida glabrata ATCC 22019 and clinical isolate were used in the study, which were obtained from the culture collection of the Department of Oral Microbiology, Faculty of Dentistry, Mahidol University.

2.3 Antifungal activity testing

The antifungal property of *C. formosum* gum was determined by disk diffusion method then further investigated for the minimum inhibitory concentration (MIC).

Overnight broth cultures of the test microorganisms were prepared in Brain Heart Infusion broth (Difco, USA), buffered with phosphate buffer saline (PBS, pH 7.4) to yield a concentration of approximately 1.5 x 10^8 CFU/mL. Paper disks (Whatmann International, UK) of 6 mm diameter, were placed on the inoculated agar surfaces and impregnated with 20 μL of the gum. For determination of MIC, the gum was dissolved in dimethyl sulfoxide (DMSO, 50% v/v). Two-fold dilution series of the gum were tested against the starting inoculums of 1 x 10^8 CFU/mL. The vehicle (50% DMSO, v/v) was used as a negative control and nystatin (100,000 units/mL, InterThai Pharmaceutical, Thailand) as a positive control

[a] sroisiri.tha@mahidol.ac.th Corresponding author, [b]boonyanit.tha@mahidol.ac.th, [c]passiri.nis@mahidol.ac.th

for growth. The MIC was defined as the lowest concentration of gum that restricted growth to a level lower than 0.05 at 600 nm (no visible growth).

2.4 Determination of α-mangostin content

C. formosum gum (50 mg) was diluted in methanol to obtain a final concentration of 1 mg/mL. The solution was filtered through 0.45 μm membrane filter prior to analysis. For standard solution preparation, a stock solution of α-mangostin reference standard was prepared to obtain concentrations of 200, 100, 50, 25 and 10 μg/mL.

The validated HPLC method was performed on Water 600 Controller, 2996 Photodiode Array Detector, a rheodyne injector fitted with a 20 μL loop and 717 Plus Auto-sampler. An XTerra® RP18 (3.9 mm x 150 mm, 5 μm particle size) with a RP 18 guard column was used. The elution was carried out with a gradient solvent system consisting of 0.1% v/v ortho phosphoric acid (solvent A) and acetonitrile (solvent B) with a flow rate of 1 mL/min at ambient temperature. The gradient program was as follows: 70% B for 0-15 min, 70% B to 75% B in 3 min, 75% B to 80% B in 1 min, constant at 80% B for 6 min, 80% B to 70% B in 1 min with 11 min of post-run for reconditioning. The wavelength of the detector was set at 320 nm. The content of α-mangostin was calculated using its calibration curve regarding the dilution factor and was expressed as g/100 g of the gum. All the tests were performed in triplicate on three separate occasions.

3 Results

The antifungal activity of *C. formosum* gum against 2 strains of *C. glabrata* was quantitatively assessed by the presence of inhibition zones and the MIC values (Table 1). The amount of α-mangostin in the gum determined by HPLC was 4.08% (w/w).

Table 1 Mean inhibition zones(mm) and MIC values (mg/mL) of *C. formosum* gum against *C. glabrata*

C. glabrata	*C. formosum*		Nystatin	
	zone	MIC	zone	MIC
ATCC 22019	14.3	1.25	12.0	0.20
Clinical isolate	10.2	1.25	11.7	0.25

4 Discussion

Cratoxylum formosum is a small tree, 10-20 m tall, which distributes in several Asian countries. It has been known for its therapeutic use on the treatment of diarrhea, food poisoning and internal bleeding [6,7]. In dentistry, the gum obtained from the burned bark of this plant has been used extensively by hill tribe people of Thailand to paint on their tooth surfaces for the prevention of tooth and gum diseases [4,8,9]. The antimicrobial properties have been reported against many types of oral microorganisms including bacteria related to tooth decay (*Streptococcus mutans*) [4] and periodontal diseases (*Porphyromonas gingivalis, Prevotella intermedia, Actinobacillus actinimycetemcomitans* and *Fusobacterium nucleatum*) [10] as well as candida (*Candida albicans*) [5]. In the present study, the gum extracted from this plant showed moderate antifungal activity against *C. glabrata* (MIC was 1.25 mg/mL) which is the second most prevalent human yeast pathogen after *C. albicans*. The effect seems less potent compared to *C. albicans* (MICs of *C. albicans* were 0.5-1.25 mg/mL). This may be due to the differences in their metabolic pathways and resistance mechanisms between these two pathogenic candida species.

In the past, *C. glabrata* was considered a relatively non-pathogenic saprophyte of the normal flora in healthy humans and was not readily associated with serious infection. However, *C. glabrata* can rapidly disseminate throughout the body and infection with this species is associated with a high mortality rate [11]. Nowadays, *C. glabrata* is of added concern because of its inherent resistance to certain antifungal agents. Data from the study of Baddley *et al.*[12] indicated that yeast species with high MICs obtained from patients with candida infection are associated with lower success rates and higher mortality than those with low or susceptible MICs [13]. Furthermore, antifungal resistance has consequences in terms of elevated MICs that are associated with poorer outcomes and breakthrough infections during antifungal treatment and prophylaxis.

Considering the toxicity of *C. formosum* gum, mildly cytotoxicity or grade 2 cytotoxicity has been demonstrated in mouse fibroblast cell line (L929) at the concentration of 1.25-12.5 mg/mL [5]. However, the acceptable biocompatibility of the agent has been set as grade 2 or lower [14].

The active components present in the plant products, xanthones are the prominent bioactive compounds in the bark or exuded gum of trees belonging to the family Clusiaceae, of which *Garcinia mangostana* L. and *Cratoxylum* spp. were traditionally employed for painting in the oral cavity [8,9,15]. Many studies have shown α-mangostin to exert the most potent antimicrobial activity against both bacteria and fungi, among all the xanthone compounds [16,17]. This study also focused on the quantification of α-mangostin and it was revealed that the amount of this xanthone derivative present in the gum was 4.08% w/w. Compared with that demonstrated by Kaomongkolgit *et al.* [17], the yield of α-mangostin from the dried pericarp of mangosteens (*Garcinia mangostana*) was approximately 10 times (0.4% w/w) less than that from *C. formosum*.

In conclusion, *C. formosum* gum showed anticandidal property in inhibiting the growth of *C. glabrata* with the MIC in acceptable biocompatibility range. Based on our data, it can be suggested that *C. formosum* gum may be a candidate for antimicrobial action against oral yeast. However, further studies should be conducted to clarify the mechanisms of action followed by clinical studies to establish the safe profile of this gum for clinical use.

Acknowledgement

The authors would like to thank the Faculty of Dentistry, Mahidol University for financial support (Grant no. 0517.0321) and the Thailand Institute of Scientific and Technological Research for advice on plant material preparation.

References

1. J Perlroth, B Choi, B Spellberg. Nosocomial fungal infections: epidemiology, diagnosis, and treatment. Med Mycol. **45[4]**, 321-46 (2007)
2. Z Jandric, C Shuller. Stress response in *Candida glabrata*: pieces of a fragmented picture. Future Microbiol. **6[12]**, 1475-84 (2011)
3. GL Tayanin, D Bratthall. Black teeth: beauty or caries prevention? Practice and beliefs of the Kammu people. Community Dent Oral Epidemiol. **34[2]**, 81-6 (2006)
4. T Suddhasthira, S Thaweboon, N Dendoung, B Thaweboon, S Dechkunakorn. Antimicrobial activity of *Cratoxylum formosum* on *Streptococcus mutans*. Southeast Asian J Trop Med Public Health. **37[6]**, 1156-9 (2006)
5. S Thaweboon, B Thaweboon, S Dechkunakorn, P Nisalak, R Kaypetch. Anticandidal activity of *Cratoxylum formosum* gum and its cytotoxicity. Advanced Materials Research. **974**, 394-7 (2014)
6. EF Anderson. Ethnobotany of hill tribes of northern Thailand. II. Lahu medicinal plants. Adv Econ Bot. **40[4]**, 442-50 (1986)
7. PW Grosvenor, PK Gothard, NC McWilliam, A Supriono, DO Gray. Medicinal plants from Riau province, Sumatra, Indonesia. Part 1: Uses J Ethnopharmacol. **45[2]**, 75-95 (1995)
8. TZ Zumbroich. Teeth as black as a bumble bee's wings': The ethnobotany of teeth blackening in Southeast Asia. Ethnobot Res Appl. **7[13]**, 381-98 (2009)
9. TZ Zumbroich. To Strengthen the Teeth and Harden the Gums-Teeth blackening as medical practice in Asia, Micronesia and Melanesia. Ethnobot Res Appl. **9[20]**, 7-13 (2011)
10. J Kuvatanasuchati, S Laphookhie, P Rodanant. Antimicrobial activity against periodontopathic bacteria and cytotoxic study of *Cratoxylum formosum* and *Clausena lansium*. J Med Plant Res. **5[25]**, 5988-92(2011)
11. PL Fidel, JA Vazquez, JD Sobel. *Candida glabrata*: review of epidemiology, pathogenesis, and clinical disease with comparison to *C. albicans*. Clin Microbiol Rev. **12[1]**, 80-96 (1999)
12. JW Baddley, M Patel, SM Bhavnani, SA Moser, DR Andes. Association of fluconazole pharmacodynamics with mortality in patients with candidemia. Antimicrob Agents Chemother. **52[9]**, 3022-8 (2008)
13. MA Pfaller. Antifungal drug resistance: mechanisms, epidemiology, and consequences for treatment. Am J Med. **125[1]**, S3-13 (2012)
14. RB Miranda, SR Fidel, A Boller. M L929 cell response to root perforation repair cements: an in vitro cytotoxicity assay. Braz Dent J. **20[1]**,22-6 (2009)
15. S Boonsri, C Karalai, C Ponglimanont, A Kanjana-opas, K Chantrapromma. Antibacterial and cytotoxic xanthones from the roots of Cratoxylum formosum. Phytochemistry. **67[7]**, 723-7 (2006)
16. PT Nguyen, RE Marquis. Antimicrobial actions of α-mangostin against oral streptococci. Can J Microbiol. **57[3]**, 217-25 (2011)
17. R Kaomongkolgit, K Jamdee, Chaisomboon N. Antifungal activity of alpha-mangostin against *Candida albicans*. J Oral Sci. **51[3]**, 401-6 (2009)

MATEC Web of Conferences **65**, 03005 (2016) DOI: 10.1051/matecconf/20166503005
ICNSCM 2016

Study on the Motion state of Powdery materials in Dense-phase Pneumatic Conveying Pipe

Zhihua Li[a], Guangpeng Liu , Nan Guo and Feipeng Liu

College of Mechanical and Electrical Engineering, Qingdao University of Science & Technology, Qingdao 266061, China

Abstract. Using the method of computational fluid dynamics, a model is created about powdery materials flowing in the dense-phase pneumatic conveying pipe. the motion state powdery materials flowing in the horizontal and vertical pipe is simulated. It is found that in the horizontal pipe the powdery materials represent the flow of dune-like state, and continuously move forward in this form, the volume fraction of powdery materials at the pipe's bottom is large, the velocity is low; In the vertical pipe, the columnar solid-plug can form and scatter continuously, and the motion state is closer to fluidization.

Introduction

Pneumatic conveying technology is using compressed gas as conveying power in a sealed pipe to convey granular materials. It is widely used to convey powdery materials such as powdery materials in rubber industry [1]. Pneumatic conveying technology has been developed from the dilute-phase conveying to the dense-phase conveying. Compared with the dilute-phase conveying, the dense-phase conveying has many advantages such as less energy consumption, large conveying capacity and low breakage rate, but there are some problems such as complex flow form, poor conveying stability and high blockage rate of conveying pipe [2]. In order to provide more precise and powerful references for the research and design of the dense-phase pneumatic conveying system, it is necessary to carry out numerical simulation and analysis of motion state of powdery materials in the horizontal and vertical conveying pipe[3].

1.The theoretical basis and establishment of mathematical model

The Fluent software in the CFD (Computational Fluid Dynamics) software package is used as the platform to simulate the motion state of powdery materials in dense-phase pneumatic conveying pipe. The problem of multiphase flow usually has two solving methods: the Euler-Euler method and the Euler-Lagrange method. The Euler-Euler method is to solve the N-S (Navier-Stokes) equation for the continuous phase fluid in the Eulerian framework, and to solve the particle phase conservation equation for the particle phase in the Eulerian framework; The Euler-Lagrange method is to solve the N-S (Navier-Stokes) equation for the continuous phase fluid in the Eulerian framework, but to solve the particle trajectory equation for the particle phase in the Lagrange framework.

1.1 The theoretical basis

The Euler-Euler model is used to calculate the gas-solid phase flow. Its basic idea is to treat the particles as pseudo fluid. The model has its own convection term, diffusion term, production term and dissipation term. In the Fluent software, there are three kinds of Euler-Euler multiphase flow model, which are VOF model, Mixture model and Eulerian model. The Eulerian model is available to simulate the particle motion state for the dense-phase pneumatic conveying of powdery materials.

Eulerian model can solve the momentum equation and continuity equation for each phase, realize coupling through pressure and interphase exchange coefficient, and obtain the flow characteristics of particles through using the molecular motion theory for granular flow. For dense-phase pneumatic conveying, the coupling between particles and gas phase is two-way, and there is a coupling between particle pressure and particle viscous stress caused by the collision between particles. The basic control equations of Eulerian model are as follows.

Continuous equation:

$$\frac{\partial \rho}{\partial t} + \nabla \cdot (\rho \vec{v}) = 0 \tag{1}$$

Momentum equation:

$$\frac{\partial (\rho \vec{v})}{\partial t} + \nabla \cdot (\rho \vec{v}\vec{v}) = -\nabla \cdot p + \nabla \cdot \vec{\tau} + \vec{F} \tag{2}$$

[a] Corresponding author: lizh@mesnac.com

Where, ρ is the density of powdery materials, kg/m^3; t is the conveying time, s; $\vec{v}$ is the conveying velocity vector of materials, m/s; p is the pressure on micro unit of fluid, Pa; $\vec{F}$ is the external body force on the micro unit, N/m^3; $\vec{\tau}$ is the viscous stress on the surface of micro unit, Pa.

1.2 The mathematical model

The model is established in the pre-processing software named Gambit. The diameter of the pneumatic conveying pipe is 150 mm. The velocity-inlet boundary condition is used as the powdery material (carbon black) inlet boundary and gas inlet boundary, and the pressure-outlet boundary condition is used as the outlet boundary. Because the pneumatic conveying pipe is slender, and the length of computational domain size is large, the author intends to use the two-dimensional double-precision solver. The standard κ-ε double equation turbulence model is used as the turbulence model. The pressure coupling equations are solved by SIMPLE (Semi-Implicit Method for Pressure Linked Equations, the half implicit method of the pressure coupling equations), which is widely used for the calculation of flow field in computational fluid mechanics. The second-order upwind is used for the difference scheme.

2 Simulation and analysis

2.1 The parameters of dense-phase pneumatic conveying system

Table 1. shows the parameters of pneumatic conveying system in the simulation process.

Table 1. The parameters of dense-phase pneumatic conveying system

Parameters of the system	The value of parameters
The stacking density of carbon black[kg/m^3]	375
The conveying capacity[t/h]	8
The conveying gas volume[m^3/h]	280
The conveying gas velocity[m/s]	10
Solid-gas ratio	22
The diameter of conveying pipe[mm]	150
The distance of horizontal conveying[m]	3
The distance of vertical conveying[m]	15

2.2 The horizontal conveying pipe

The phase diagram is a set of curves to describe the relationship between superficial gas velocity ug and unit length of pipe pressure drop Δp_L through the solid mass flow q_m. It can accurately describe the flow characteristics of pneumatic conveying, and intuitively present the relationship between pressure loss and superficial gas velocity. It also can be used on the research of flow form division, stability analysis and optimization of parameters [4]. The typical phase diagram of Zenz in horizontal pipe is shown in figure 1.

In figure 1, when the superficial gas velocity ug at point A and larger, the flow form is dilute phase suspension flow. The influence of materials gravity is small, and the suspension force acted on powdery materials mainly comes from the vertical velocity component of turbulent gas flow and the bounce force when powdery materials collide with pipe wall. When the superficial gas velocity reduces to point B, a portion of particles begin to settle down from the suspension flow. And there is a concentration gradient of powdery materials in the pipe cross section, which forms the stratified flow. When the superficial gas velocity continues reducing to point C, the solid phase starts to separate from gas phase, and particles start to slide along the bottom of the pipe at a low speed in the role of air drag force and collision, which forms the moving bed. Curve CEF represents the connection line of the minimum pressure drop, as well as the dividing line of stable flow and unstable flow. Continue to reduce the superficial gas velocity to point D, and the powdery materials move forward continuously in dune-like state with the accumulation of particles. The powdery materials that come into the pipe later repeat the process of settlement, showing the next dune-like state. So the forward motion state of a section of dune-like materials has formed in the whole horizontal pipe. At present, the particle-particle interactions and the interaction of particles and pipe wall have the biggest effect on the motion state.The simulation about the motion state of powdery materials in horizontal conveying pipe at different times is shown in figure 2.

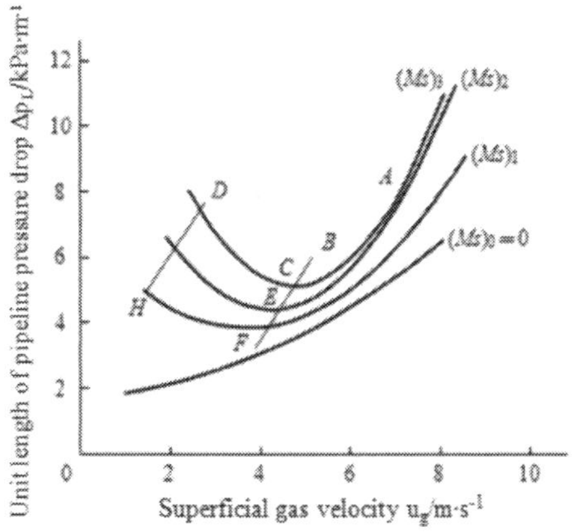

Figure 1. The typical phase diagram of Zenz in horizontal pipe

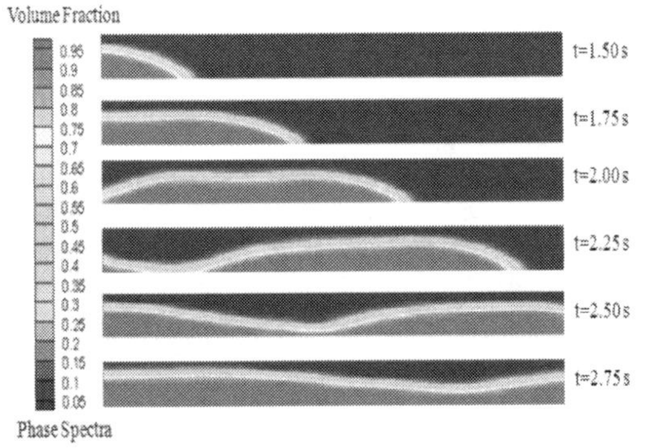

Figure 2. The motion state of the powdery material in horizontal conveying pipe at different times

Figure 3. and figure 4. respectively show the volume fraction of the powdery material and the distribution of velocity along the pipe diameter (Y-axis) in a certain cross section of the horizontal conveying pipe. In the figures, Y = 0 m represents the bottom of the pipe and Y = 0.15 m represents the top of the pipe.

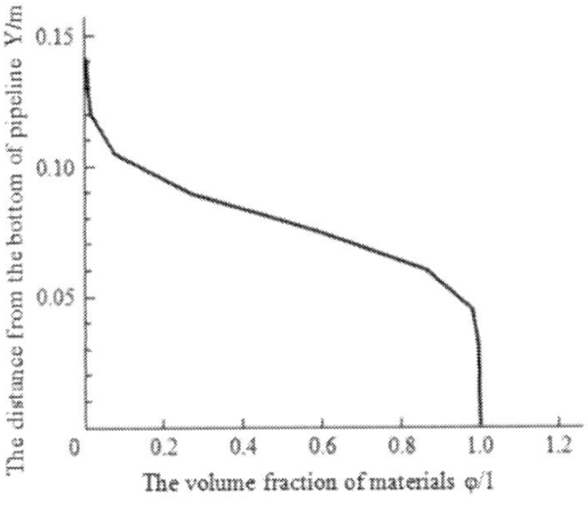

Figure 3. The volume fraction of the powdery material changes along the pipe diameter direction

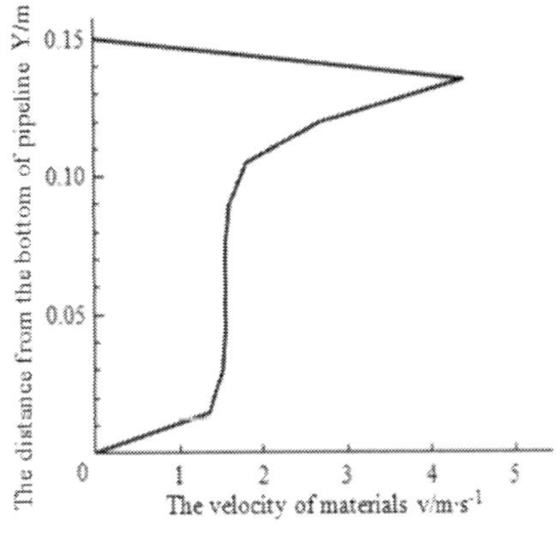

Figure 4. The velocity of the powdery material changes along the pipe diameter direction

The comprehensive analysis about figure 2., figure 3. and figure 4. shows that the powdery material particles mostly concentrate in the bottom of the horizontal conveying pipe, the void ratio of the powdery material is small, and the velocity of the powdery material is relatively low. The volume fraction of the powdery material at the top of the horizontal pipe is small, the void ratio of the powdery material is high, and the velocity of the powdery material is relatively high. Solid-plug moves forward in the form of wave, along with the displacement movement of particles in solid-plug and accumulation layer. The velocity of the powdery material in the pipe wall is zero, the more close to the pipe wall, the greater velocity gradient of the powdery material will be. In figure 3, the accumulation layer thickness of the powdery material can be determined by the height of curve that the volume fraction of the powdery material is close to 1. Accumulation layer thickness will decrease with the increase of superficial velocity. The velocity of the powdery material in the accumulation layer is very low, and it moves forward with the entrainment effect caused by the movement of the dune-like solid-plug. The thickness of accumulation layer changes along with the conveying distance, the more close to the end of the conveying pipe, the smaller accumulation layer thickness will be. The simulation results are consistent with the experimental results [5].

2.3 The vertical conveying pipe

The relationship between unit length of pipe pressure drop Δp_L and superficial gas velocity ug in the vertical conveying pipe for different conveying capacity is shown in figure 5 [6].

As figure 5 shows, the relationship about the volume flow of powdery materials is $Q^0 < Q^1 < Q^2 < Q^3$. Segment AB represents the relationship between unit length of pipe pressure drop ΔpL and superficial gas velocity ug under the condition that the volume flow of powdery materials $Q^0 = 0$. Taking curve Q^1 as an example, when the superficial velocity is high, the volume fraction of powdery materials is relatively low. In this case, the powdery material is in dispersed state, that is, the dilute phase conveying. In the CD stage, with the reduction of superficial gas velocity, the solid-gas mixture ratio increases gradually, the material gravity pressure drop Δp_g increases, but the overall pressure drop Δp decreases. This shows that the pressure drop Δp_f generated by pipe wall friction reduces much larger than the increase in the pressure drop Δp_g generated by the powdery material gravity. In the DE stage, with the continuous reduction of superficial gas velocity, the overall pressure drop Δp decreases gradually and reaches the lowest values, but the proportion of the pressure drop Δp_g generated by the powdery material gravity is large. In the EF stage, with the further reduction of superficial gas velocity, the solid-gas mixture ratio increases further, the pressure drop Δp_g generated by the powdery material gravity plays a dominant role in the pipe. The overall pressure drop Δp increases, and the volume fraction of powdery materials in the pipe is relatively high. In this case, the powdery material is in aggregation state, that is, the dense phase conveying. When the superficial gas velocity is lower than its value in point F, the relative velocity between gas and particles cannot make the powdery material remains in suspension, which will produce a packed bed flow or moving bed flow.

The simulation about the motion state of the powdery material in vertical conveying pipe at different times in the EF stage is shown in figure 6. Because of the adhesion among material particles, the particles gather together and move upward in the role of suspension force generated by the upward flow of gas. When the solid-plug moves upward, it undertakes the falling materials from the previous solid-plug, and scatters corresponding amount of materials at the end of the solid-plug. So the solid-plug bed is similar to the fluidized state. The simulation results are consistent with the experimental results [5].

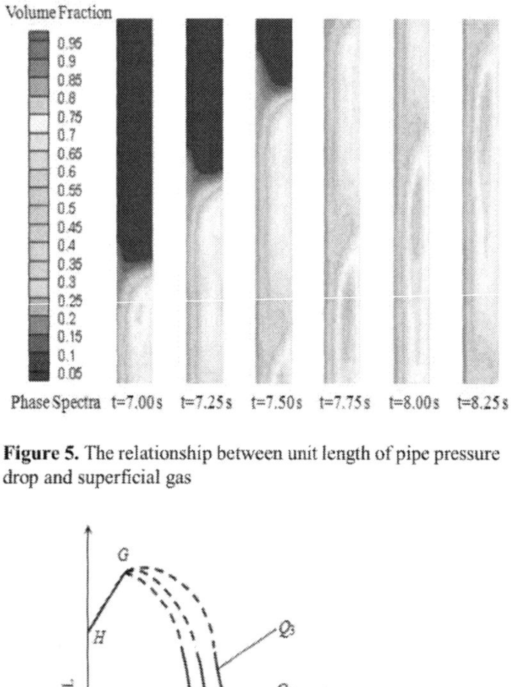

Figure 5. The relationship between unit length of pipe pressure drop and superficial gas

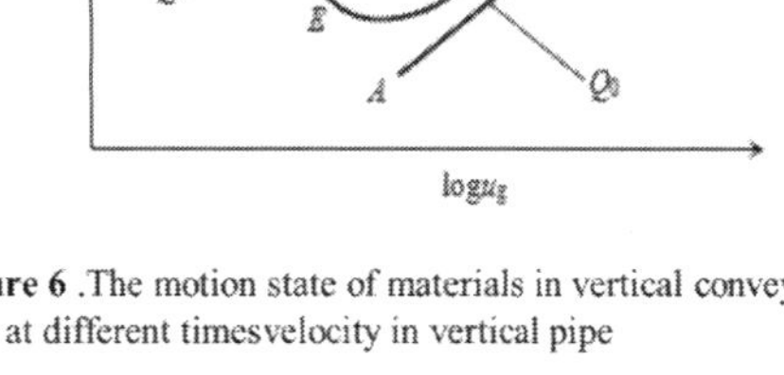

Figure 6 .The motion state of materials in vertical conveying pipe at different timesvelocity in vertical pipe

3 Conclusion

The following conclusions can be obtained by using Fluent software to simulate the flow process of powdery materials in dense-phase pneumatic conveying horizontal and vertical pipes.

(1) In horizontal pipe conveying process, powdery materials can represent the flow of dune-like state, and continuously move forward in this form. Powdery materials which are deposited at the bottom of pipe can form the accumulation layer. The larger the superficial gas velocity is, the smaller the thickness of accumulation layer will be. The volume fraction of powdery materials at the bottom of pipe is large, so the velocity is low, and the volume fraction of powdery materials at the top of pipe is small, so the velocity is high.

(2) In vertical pipe conveying process, because of the two-way role of suspension force generated by the upward flow of gas and the gravity of powdery materials, the columnar solid-plug can form and scatter continuously, showing the state of fluidization.

(3) The simulation results are consistent with the actual pneumatic conveying experimental results. This shows that the numerical simulation method can be used to research the dense-phase pneumatic conveying mechanism and the optimization design of pneumatic conveying system.

References

1. S.L.Chen,X.D.Yang, Q. Shi, Traffic information and security,30,74 (2012)
2. L.Yang,Y.H.Xie,*Pneumatic Conveying Engineering*, (China Machine Press, Beijing,2006)
3. Z.H. Li,X.C. Li,G.P. Liu, Rubber Industry,**62**,496 (2015)
4. Q.M. Meng, Y. Zhou , X.P. Chen, et al, Boiler Technology, **42** ,4(2011)
5. Z.H. Li, Y.Q.Liu ,P. Xia, Hoisting and Conveying Machinery, **87**,75 (2009)
6. W.X. Peng ,Y.Q. Wu , *Pneumatic Conveying and Transporting in Workshop* (China Measure Press, Beijing,2009)

DOI: 10.1051/matecconf/20166503006

Analysis and Study on Performance of a New Integrated Dust Precipitator

Zhihua Li[a], Fengxue Dai , Feipeng Liu and Nan Guo

College of Mechanical and Electrical Engineering, Qingdao University of Science & Technology, Qingdao 266061, China

Abstract. The dust precipitator is integrated with cyclone and bag filter, With the aid of a mathematical model this paper proposed. the flow characteristics of dust-gas inside the new dust precipitator is studied. It is concluded that the new integrated dust precipitator combines the advantages of the cyclone and bag filter, avoids the violent fluctuation to bags when working, and raises the filtrating efficiency.

Introduction

The dust precipitator is used to separate and collect dust from dust-gas flow. It's the key equipment for a dust removal system. There are many types of precipitators used in industry. The cyclone, which is one of the most widely used equipment, has no moving parts inside [1]. But with the help of the dust-gas flow centrifugal force, only some bigger solid granules maybe separated from the dust-gas. Cyclone is often used as the first step equipment in the dust removal system. Bag filter separates the dust from the dust-gas by using organic fiber or inorganic fiber as filter bags. Bag filter is suitable as the second step equipment in the dust removal system, especially has comparative advantages for removing small solid particles. A new type of dust precipitator, integrating the two equipments, is put forward in this paper. The new dust precipitator not only simplifies the dust removal system, but also saves installed space, raises the filtrating efficiency.

1 The structure and working principle

The new dust precipitator consists of cyclone unit and filtrating bag unit. The structure sketch is shown in Figure 1. The cyclone unit includes dust-gas inlet, external cyclone shell, internal cyclone shell and ash outlet. The filtrating bag unit includes clean gas outlet, compressed gas tank, electromagnetic pulse valve, filter bag, bag framework, perforated plate, casing of filter and pulse control instrument. The two units are directly connected up and down with bolts in its entirety. Dust-gas goes through the dust gas inlet into the cyclone unit firstly. Inside the unit, some bigger solid granules are separated from the dust-gas. Then the gas enters to the filtrating bag unit, through the bags' filtrating effect, the residual solid granules are separated from the gas. After the two filtrating process, the much more clean gas vents to the gas outlet. And the ash discharges through the ash valve. Thus, one set of this new dust precipitator may achieve two steps of dust removal, and instead of two dust removal equipments such as one set of cyclone and one set of bag filter.

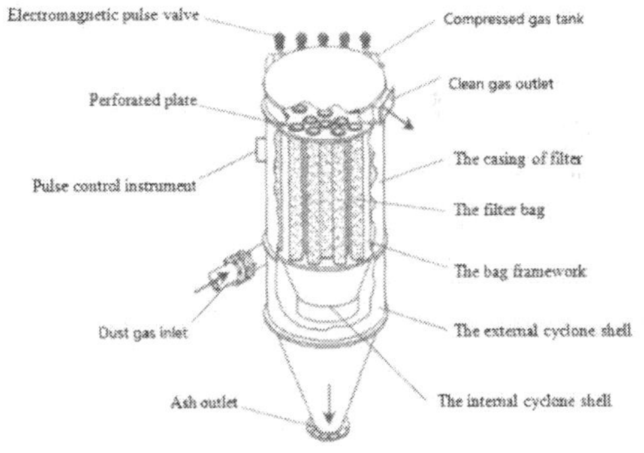

Figure1. The structure sketch about the new dust precipitator

2 The numerical simulation

2.1 The establishment of mathematical model

The dust gas inlet diameter of the new dust precipitator is d=0.14 m, the velocity is u=12 m/s. If the ambient temperature is 20 ℃ , the gas dynamic viscosity is μ_g=17,9×10-6 Pa·s, the density is ρ_g=1.205 kg/m^3. So, the Reynolds number of the flow field inside the new dust precipitator is

$$\mathrm{Re} = \rho_g ud / \mu_g = 1.205 \times 12 \times 0.14 / (17.9 \times 10^{-6}) = 113095 > 12000 \quad (1)$$

[a] Corresponding author: lizh@mesnac.com

According to the calculation result, the conclusion is known that the state of the flow field being located next to the gas flow inlet of the new dust precipitator is turbulent. As RSM model reflects the characteristic of the turbulence anisotropy, and has been widely used to simulate the gas-solid two-phase flow inside the conventional filters and achieved similar result compared with the actual state [2]. Furthermore, when the new dust precipitator works, three-dimensional strong rotary flow will comes into being inside it, and has the obvious characteristic of the turbulence anisotropy. So, if the whole flow can be regarded as unsteady compressible, the new dust precipitator may be simulated and calculated with RSM model according to the hydromechanics software Fluent [3]. The control equations are as follows.

Continuity equation:

$$div\overline{V} = 0 \tag{2}$$

Momentum equation:

$$\frac{\partial \overline{u}}{\partial t} + \text{div}(\overline{u}\overline{V}) = -\frac{1}{\rho}\frac{\partial \overline{p}}{\partial x} + \nu\text{div}(\text{grad}\overline{u}) - [\frac{\partial \overline{u'^2}}{\partial x} + \frac{\partial \overline{u'v'}}{\partial y} + \frac{\partial \overline{u'w'}}{\partial z}] \tag{3}$$

$$\frac{\partial \overline{v}}{\partial t} + \text{div}(\overline{v}\overline{V}) = -\frac{1}{\rho}\frac{\partial \overline{p}}{\partial y} + \nu\text{div}(\text{grad}\overline{v}) - [\frac{\partial \overline{u'v'}}{\partial x} + \frac{\partial \overline{v'^2}}{\partial y} + \frac{\partial \overline{v'w'}}{\partial z}] \tag{4}$$

$$\frac{\partial \overline{w}}{\partial t} + \text{div}(\overline{w}\overline{V}) = -\frac{1}{\rho}\frac{\partial \overline{p}}{\partial z} + \nu\text{div}(\text{grad}\overline{w}) - [\frac{\partial \overline{u'w'}}{\partial x} + \frac{\partial \overline{v'w'}}{\partial y} + \frac{\partial \overline{w'^2}}{\partial z}] \tag{5}$$

Where, $\overline{V}$ is the velocity vector, m/s; $\overline{u}$ $\overline{v}$, $\overline{w}$ are the subordinate and average velocity vectors in the direction of the x, y, z, m/s; u', v', w' are the pulse velocity in the direction of x, y, z, m/s.

The turbulent kinetic energy k equation and the dissipation rate ε equation are as follows:

$$\frac{\partial(\rho k)}{\partial t} + \frac{\partial(\rho k u_i)}{\partial x_i} = \frac{\partial}{\partial x_j}[(\mu + \frac{u_j}{\sigma_k})\frac{\partial k}{\partial x_j}] + \frac{1}{2}p_{ij} - \rho\varepsilon \tag{6}$$

$$\frac{\partial(\rho\varepsilon)}{\partial t} + \frac{\partial(\rho\varepsilon u_i)}{\partial x_i} = \frac{\partial}{\partial x_j}[(\mu + \frac{\mu_t}{\sigma_\varepsilon})\frac{\partial\varepsilon}{\partial x_j}] + \frac{1}{2}C_{1\varepsilon}p_{ij} - C_{2\varepsilon}\rho\frac{\varepsilon^2}{k} \tag{7}$$

Where, ρ is the fluid density, kg/m^3; k is the turbulent kinetic energy, m^2/s^2; μ is the dynamic viscosity, Pa·s; μ_t is the turbulent viscosity, $\mu_t=\rho C_u k^2/\varepsilon$, Pa·s; p_{ij} is the shear stress production, 1; ε is the turbulent dissipation rate, m^2/s^3; C_u, $C_{1\varepsilon}$, $C_{2\varepsilon}$ are experience constants, C_u=0.09, $C_{1\varepsilon}$=1.44, $C_{2\varepsilon}$=1.92; σ_k, σ_ε are the turbulent Bronte numbers of turbulent kinetic energy k and turbulent dissipation rate ε, σ_k=0.82, σ_ε=1.3.

2.2 The filter's geometrical parameters

The dust-gas inlet is designed in the middle of the new dust precipitator. For the new dust precipitator, the total height is 2600mm, and the height of the clean gas chamber is 200mm. The diameter of the cylinder is 880mm, and its height is 1600mm. The height of the cone is 800mm, and its bottom diameter is 200mm. The diameter of the dust gas inlet is 140mm. The clean gas outlet is situated in the clean gas chamber, its width is 400mm, height is 160mm, nineteen filter bags are arranged along the radial direction, and the filter bag specification is Φ130 mm×1200 mm.

2.3 The boundary and initial conditions

For the new dust precipitator, the velocity is selected as the inlet boundary condition, and the free outflow is selected as the outlet boundary condition. It is utilized the bag's film property and chose the porous jump boundary condition to simulate the bag's filtrating [4]. The turbulent intensity is I=0.16$Re^{-1/8}$.

The input boundary condition values using for simulation and calculation about the model of the new dust precipitator are shown in Table 1.

Table 1. The input boundary conditions

The input boundary conditions	The value of boundary conditions
The inlet diameter[m]	0.14
The inlet velocity[m/s]	12
The turbulence intensity[%]	4
The surface permeability[m^2]	1×10^5
The thickness of perforated plate[mm]	3
The pressure step coefficient[m^{-1}]	0

2.4 The simulation results and analysis

2.4.1 The simulation results

The velocity vector and pressure of the new dust precipitator taking from simulation are as shown in Figure 2 and Figure 3.

From Figure 2 and Figure 3 , when dust-gas goes into the cyclone unit at the velocity of 12 m/s, a strong swirl appears in its cone, and the more distance away from the cone centre, the more high velocity and pressure. On the circumference near the inlet the velocity and pressure reach to the maximum, but near the center part fall significantly.

By means of the strong swirl effect, coarse particles and high density particles from the dust-gas directly drop to the cone's bottom, which reduces the impact on the bags. Thus, the bag life is extended and the injection cleaning frequency may be reduced so as to saving energy. At the same time, the most critical, it significantly raises the dust removal efficiency.

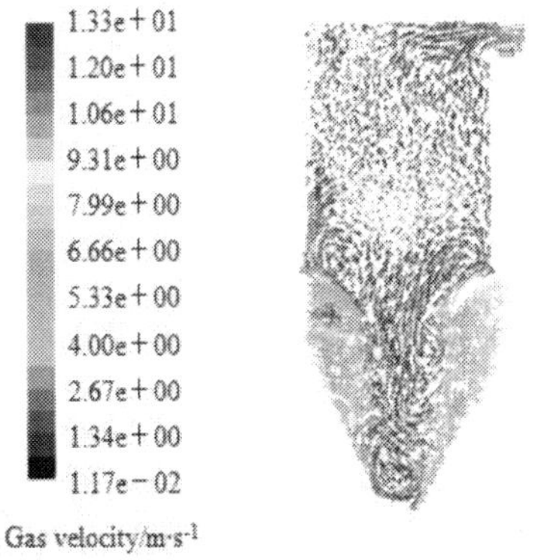

(a) The velocity vector of vertical section

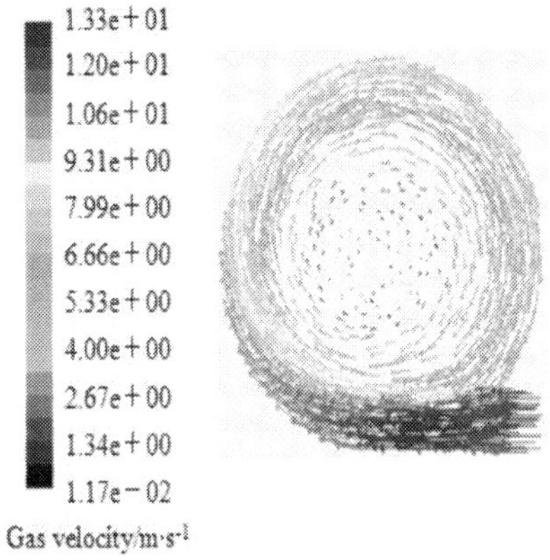

(b) The velocity vector of cross section near inlet

Figure 2. The velocity vector about the new dust precipitator

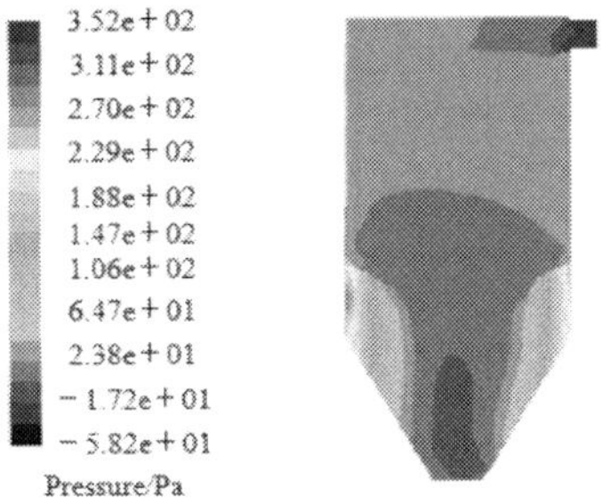

(a) The pressure cloud of the vertical section

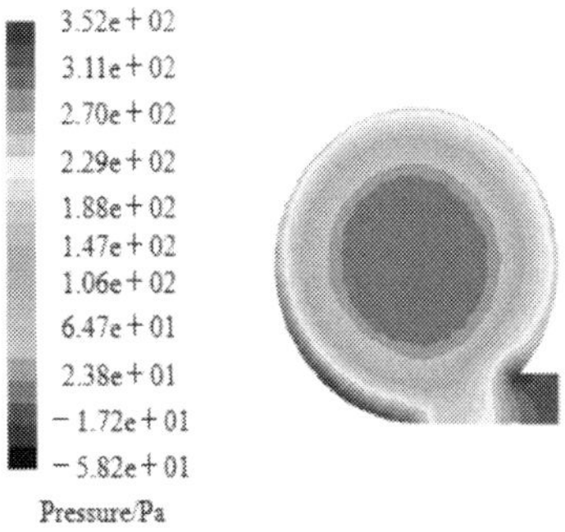

(b) The pressure cloud of the cross section near inlet

Figure 3. The pressure cloud about the new dust precipitator

2.4.2 The internal gas distribution of the filter

The layout of the filtering bags in the new dust precipitator is shown in Figure 4. The gassy mass flow rate through each filter bag is shown in Table 2.

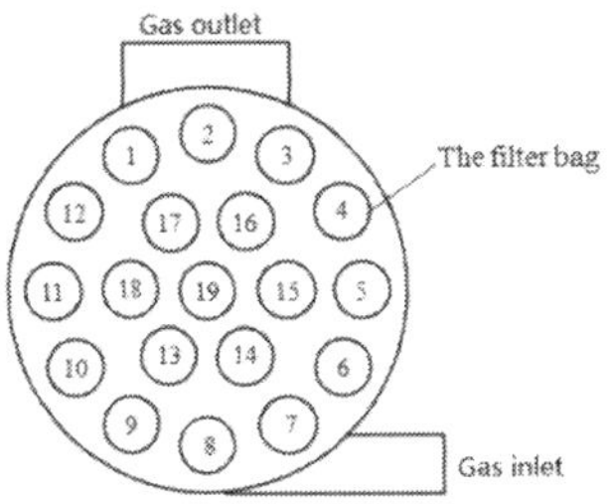

Figure 4. The layout of the filter bags

Table 2. The gassy mass flow rate through each filter bag

Serial number	Gassy mass flow rate
1	0.143549350
2	0.178450410
3	0.142806680
4	0.090841241
5	0.065733500
6	0.048651427
7	0.038259007
8	0.031632427
9	0.030742709
10	0.046368368
11	0.073229834
12	0.097862132
13	0.043250829
14	0.043056201
15	0.060569424
16	0.088580698
17	0.091376156
18	0.063441195

19	0.060418278

By analyzing the corresponding data from Figure 4 and Table 2, it is known that the bags near the gas outlet of the filter number 1, 2 and 3 passed through the maximum amount of gas, the bags away from the gas outlet number 7, 8, 9 passed through the minimum amount of gas. This is because, when the dust-gas goes into the inlet of the new dust precipitator, it would rotate downward along the surface of the cyclone unit to finish the first step dust removal. Then, the dust-gas rotates upward along the axis into the bag filter unit. Due to the gas moving toward the gas outlet, making the flow rate for each bag fluctuating but small. Of course, by optimizing the layout of filter bags and fitting a guide plate, the gas distribution will be more uniform. So, the structure of the new dust precipitator can avoid the defect of large flow rate at the gas inlet zone for the traditional bag filter, and it cannot make the bags have violent fluctuation.

2.4.3 The analysis of residuals and mass flow rate

Through optimizing the grid, improving the set of various parameters, choosing rational initialization parameters, and controlling relaxation factor, convergence of iteration may be ensured and improved. When calculation reaches convergence, whether or not the balance about the inlet or outlet gas mass flow rate of the model is reached should be checked, and whether or not the calculations achieves convergence should be judged according to the balance of the residual history and gas mass flow rate. Through multiple iterations, the final result is near the residual preset 1×10^{-3}, the iterating is convergence.

Through Fluent calculating, it is known that, for the new dust precipitator, its inlet gas mass flow rate is 1.4401137 kg/s, outlet gas mass flow rate is 1.440114 kg/s, and the difference value is 3.0×10^{-7} kg/s. The non-balance ratio is $3.0\times10^{-7}/1.440114=2.08\times10^{-7}<0.01\%$.

This calculation verifies that gas mass flow rate goes through inlet or outlet of the filter model having come to balance and the result being accurate.

3 Conclusion

The new type of new dust precipitator combines the advantages of cyclone and bag filter, integrates the two step traditional dust removal equipments to one . Its structure is simple and compact, and the dust removal efficiency is rising.

when the dust-gas goes through the inlet into the cyclone unit of the new dust precipitator, the outside vortex takes away the particles from the wall to the bottom and removes them by ash valve. The inside vortex takes the small particles (not separated) to the filtrating bag unit. The remanent small particles are obstructed by the bags and clean gas goes through the bags to atmosphere. In the filtrating bag unit the gas velocity is significantly smaller, and the pressure is much lower. At the same time, the low dust concentration of the inside vortex may decrease the dust jet times, thus not only extends the using life of bags, but also saves energy.

The layout of each filter bag affects the amount of its processing mass flow. The numerical simulation and optimization of bag's arrangement may achieve the uniform of gas flow, and provide theoretical evidence to research and design this type of dust precipitator.

References

1. X. H. Hou, W.Y. Chen, Mining & Processing Equipment, **40**, 98(2012)
2. Y. Li, L. S. Xin, Journal of Qingdao University of Science and Technology, **31** , 177(2010)
3. S.S. Zhang, *Computational fluid dynamics and its application: the principle and application of CFD software* (Huazhong University of Science and Technology , Wuhan,2011)
4. S.H. Li, Y. Song, T. Y. Wang, et al, Journal of Northeast Dianli University, **35** , 12(2015)

MATEC Web of Conferences **65**, 03007 (2016) DOI: 10.1051/matecconf/20166503007
ICNSCM 2016

The Reliability Study of Ram BOP Seals of Temperature snubbing

Jiaqi Jin[1,a], Xianrong Wang[2,b],Yuan Liang[3,c] and Wei Zhou[4,d]

[1,3,4]Schoolof mechanical engineering, Shenyang University of Technology, Shenyang, PR China

[2]Institution of drilling tool, Liao He oilfield of CNPC, Panjin, PR China

ajinjq612@126.com, c315078020@qq.com

Abstract. In order to study the safety and reliability of the seal ram BOP, this article analyzes the structure and principles of ram BOP, draws the reason of seals failure by analyzing the forces of sealant core; uses accurate simulation on plastic core by ANSYS Software and then draws that morph mainly occurs in front of the plastic core. After the scene several simulation experiments, the damage is consistent with the results of finite element analysis.

1 Introduction

With the further development of oil and gas fields, oil and water wells maintenance have become more and more frequent . Conventional operations generally use high-density killing hydraulic well or using blowout means to reduce the well pressure to conduct operations, high-density killing fluid not only the high cost, long time, and will plug the underground water channels, affecting oil production and water injection effect[1]. If a blowout buck, will not only significantly reduce the formation energy, bring inestimable loss to follow-up production, at the same time, also can cause a lot of environmental pollution. The blowout preventer seal system is the core device of snubbing system equipment, sealing performance is directly related to the effectiveness and success of the job under pressure.

2 System function and structural principles

In the high temperature and pressure operation,using ram BOP workover equipment can solve technical problems of three areas:

[c] Corresponding author: 463494796@qq.com

a. Fast switching action, safe, reliable and easy to use, in exceptional cases, can cut off the pipeline;

b. In the process of construction operation, realized the tubing, casing annulus dynamic seals and internal clogging of column, through the manifold that shell bypass outlet connect conduct the mud circulation, well pressure operation;

c.The lower pipe column process can overcome the pressure in the well on the string of jacking force, to achieve safety and no pollution pressure construction.

The overall structure consists of housing, side door, sealing ram assembly, piston, cylinder, cylinder head, locking shaft, shields , etc.

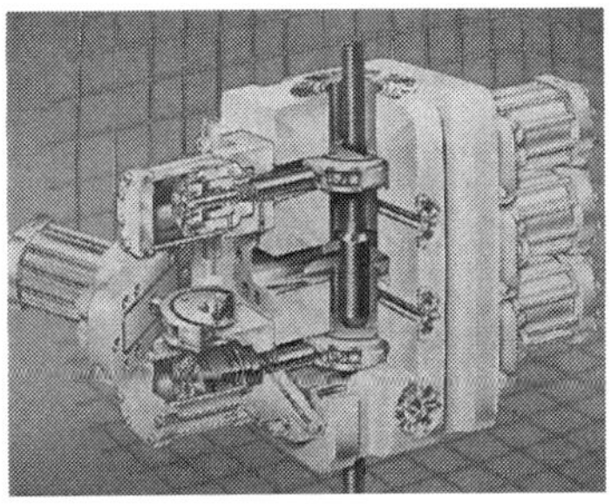

Figure 1. BOP system structure

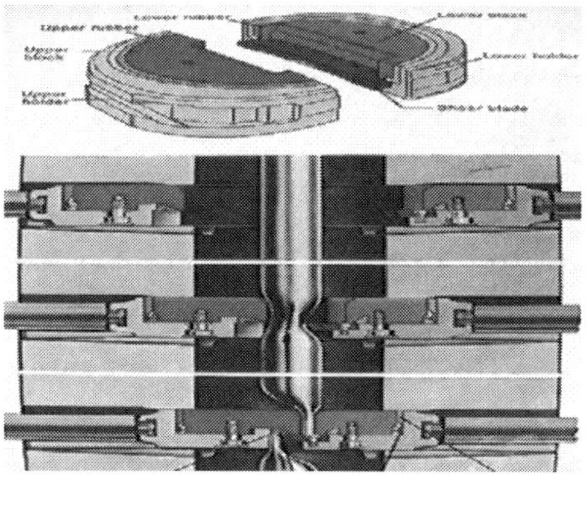

Figure 2. The three gate BOP principle diagram

Shutter switch is controlled by hydraulic valve control system, when the high-pressure oil of hydraulic control system into both sides of the hydraulic cylinder ram closed chamber,will push the piston to drive the gate shaft and both sides of flashboard assembly along the shell of a flashboard cavity respectively move to the wellhead center and achieve plugging. When the high-pressure oil of hydraulic control system into both sides of the hydraulic cylinder ram open cavity,will push the piston to drive the gate shaft and bothsides of flashboard assembly to leave the center of the well head and open the wellhead. The sealing process of the gate is divided into two steps: first, under the effect of hydraulic oil, the gate shaft push the front gate sealant core so that the front gate sealant core is extruded and deformed,the Interference between top of sealant core and the shell compress top of sealant core and then form initial sealing; second, when there is pressure in the well, well pressure pushing top of gate sealant core from gate rear make the front gate sealant core further extruded and deformed, at the same time the well pressure push the gate floating and snapping the housing sealing surface, thereby forming a reliable seal, that is the role of well pressure to help seal[2].

3 Sealant core failure analysis

When the gate rubber core sealing pipe column, the circular arc surface of rubber core maintain sufficient contact pressure with pipe column diameter, the rubber core is extruded and deformed, according to the wear delamination theory when sliding friction at low speed, the total amount of wear between hard surface and soft surface is proportional to positive pressure and sliding distance. Therefore, when working with pressure, The higher liquid pressure of injection wells, plastic core sealing surface more susceptible to wear, until they lose the sealing contact pressure that pressure operation condition should have, thus resulting in seal failure.

From the actual working conditions of ram BOP plugging, considering the role of well pressure on the front part of the rubber paraboloid, obviously, the actual deformation of free surfaces can be obtained according to the elastic mechanics knowledge, more accurate calculation of finite strain expressions as follows:

$$\varepsilon_x=\frac{\partial u}{\partial x}+\frac{1}{2}\left[\left(\frac{\partial u}{\partial x}\right)^2+\left(\frac{\partial v}{\partial x}\right)^2+\left(\frac{\partial w}{\partial x}\right)^2\right] \quad (1) \qquad \gamma_{xy}=\frac{\partial u}{\partial y}+\frac{\partial v}{\partial x}+\frac{\partial u}{\partial x}\frac{\partial u}{\partial y}+\frac{\partial v}{\partial x}\frac{\partial v}{\partial y}+\frac{\partial w}{\partial x}\frac{\partial w}{\partial y} \quad (4)$$

$$\varepsilon_y=\frac{\partial v}{\partial y}+\frac{1}{2}\left[\left(\frac{\partial u}{\partial y}\right)^2+\left(\frac{\partial v}{\partial y}\right)^2+\left(\frac{\partial w}{\partial y}\right)^2\right] \quad (2) \qquad \gamma_{yz}=\frac{\partial v}{\partial z}+\frac{\partial w}{\partial y}+\frac{\partial u}{\partial y}\frac{\partial u}{\partial z}+\frac{\partial v}{\partial y}\frac{\partial v}{\partial z}+\frac{\partial w}{\partial y}\frac{\partial w}{\partial z} \quad (5)$$

$$\varepsilon_z=\frac{\partial w}{\partial z}+\frac{1}{2}\left[\left(\frac{\partial u}{\partial z}\right)^2+\left(\frac{\partial v}{\partial z}\right)^2+\left(\frac{\partial w}{\partial z}\right)^2\right] \quad (3) \qquad \gamma_{zx}=\frac{\partial w}{\partial x}+\frac{\partial u}{\partial z}+\frac{\partial u}{\partial z}\frac{\partial u}{\partial x}+\frac{\partial v}{\partial z}\frac{\partial v}{\partial x}+\frac{\partial w}{\partial z}\frac{\partial w}{\partial x} \quad (6)$$

In the formula, ε_x , ε_y , ε_z and γ_{xy} , γ_{zy} , γ_{zx} represent the normal strain and shear strain respectively, u , v , w are the direction of displacement of the three coordinate axes[3]. According to the statistical theory of rubber elasticity, obtain the stress-strain special relationship of volume before and after deformation, according to the boundary condition, actual deformation of the front of the rubber can be solved in theory[4].

4 Finite element analysis

Analyzing plastic core top model and doing finite element network partitioning by Marc software[5]. In addition to the front end face of the node,z displacement is zero of other nodes; x, y, z displacement of some nodes is zero that contact with ram body arc groove; x displacement on both sides of the nodes is zero. The inner rubber core of the model is sealed with rubber material, elastic modulus: 7.8E+0.6PA, poisson's ratio: 0.47PA, the contact surface is a hexahedron consists of 19948 units, 93259 nodes.

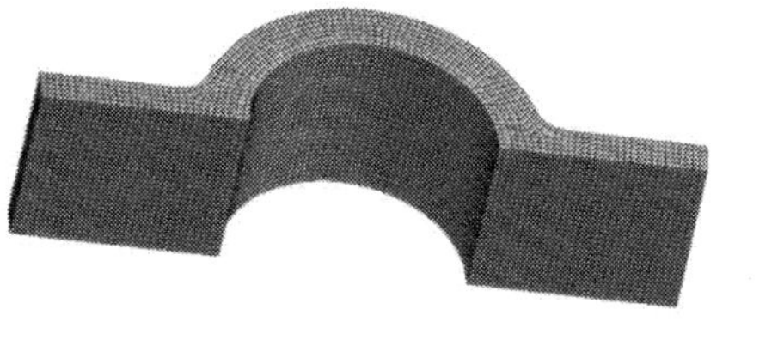

Figure 3. Seal mesh partition

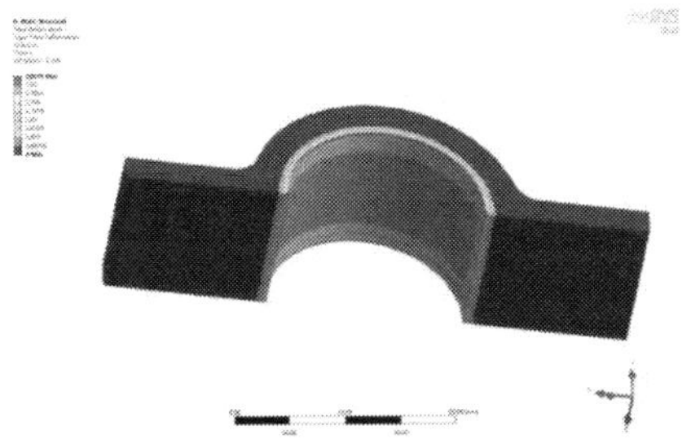

Figure 4. Seal the whole stress

As we can see from Figure 3and 4， plastic deformation occurs mainly in the front part of the core, rubber around the ribs which in the middle of the four skeleton of plastic core front portion have a greater deformation, this shows where the rubber is extruded,which is the main source of the closed drill pipe rubber core arc surface of the sealing gap. After the rubber is extruded from the skeleton,and then squeezed by string, produce z displacement,the rubber extruded due to limitations of backbone bear tensile stress besides suffering extruded tube column stress. So it is easy to make plastic core destruction.

5 Trial production of 7 "wellhead equipment simulation test

To suppress to the annulus using high-pressure pump, BOP were closed in each group, check the sealing condition, results as shown in Table 1.

Table 1. The Results of Field Experiments

	Test pressure Mpa	Frequency	Test results
Semi rams	21	3	10min drop<0.7 MPa No leakage
Full closure gate	21	3	10min drop<0.7 MPa No leakage
Shear ram	21	1	14MpaControl pressure cut 114 Insulated pipe
The ring	21	3	10min drop<0.7 MPa No leakage
Full closure gate	21	3	10min drop<0.7 MPa No leakage
Under Circular	21	3	10min drop<0.7 MPa No leakage

The above table shows: semi rams, full closure gate, shear ram, the ring, full closure gate, under circular blowout preventer system all have no leakage, at this point the rubber core still has the sealing capability.

Figure 5. Sealants core fatigue test

As can be seen from Figure 5, the transition part appear serious abrasion between front plane and circular surface of plastic core, damage is consistent with the finite element analysis results.

References

1. Yi Xin Tang. Snubbing from the drilling equipment and technology [J]. Drilling & Production Technology,1995,18(11):122-128.
2. Yu Zhang, Shu ren Zhang. New breakthrough of sealing technology:Adjustable seal [J]. Hydraulic and pneumatic seals , 1999 (6) : 42-43.
3. Bei chen Liu, Hong sen Lu, Elastic mechanics [M]. Metallurgical industry press,1979
4. P.K Kerry Frey,A.R Payne,Theory and Practice that rubber used in engineering,Chemical industry press,1985.
5. Changfeng Zhou,Jianrun Zhang. Rubber buffer finite element analysis [J]. Elastomer2004, 14 (1) : 39-42.

Fund Project:Research on Key Technology of

Truck-mounted steam engine Flooding

with hing temperature snubbing workover.

（2013020087-301）

Development and research of vacuum-plasma composite coatings for increase of operability of shearing dies

Marina A. Volosova [1], Anton E. Seleznev [1], Alexander A. Shein [1]

[1] Moscow State University of Technology "STANKIN",1 Vadkovskiy per., Moscow, 127055, Russia Federation

Abstract. The technological principles of low-temperature deposition of multilayer vacuum-plasma coatings with the increased operational properties on various types of die steels with the assistance of accelerated argon molecules beam are offered in the article. Four standard architecture of vacuum-plasma coatings - Ti/(Ti,Zr)N, Cr/CrN, TiN/TiCN/(Ti,Cr)N and Ti/(Ti,Al)N are considered. It is shown that deposition of vacuum-plasma coatings of the listed structures at the optimum modes considerably reduces intensity of wear of shearing dies therefore quality of shaped products several times increases.

Keywords: dies, wear, cutting-out, low-carbon steel, aluminum alloy, corrosion-resistant steel, carbon steel.

1 Introduction

The practice shows that the intensive wear of the working surfaces of the shearing dies substantially reduces indicators of quality of the stamped parts, increases the resistance to separation up to 20%, and the total work of separation up to 40% compared with the operation activity of the tool with sharp cutting edges. The common industrial method of reducing the wear rate of metal-working tools is the application of different coatings of the working surfaces [1, 2, 3]. In forging production the coating technologies do not have wide application, primarily due to the complicated dynamic loads imposed on the working surfaces of tools in the process of stamping, which makes inefficient the use of coatings, traditional for other metalworking areas, such as TiN, TiCN, and others.

2 Features of the wear of shearing dies and functions of the coatings

Depending on technological conditions of the shearing stamp process and on operating loads, there may be different fracture mechanisms and failure modes of the punch and the die - adhesive and abrasive wear, chipping of the work surfaces, tempering of a thin surface layer and an intense pickup of deformable material on the operating surface of the tool [4, 5].

The analysis of the results of experimental research shows that, depending on the conditions of the shearing stamp process wear resistant coatings applied on the working surface of the tool should perform the functions of:

- highly rigid surface layer that prevents abrasive wear of the working surfaces of the tool in the stamping process;
- intermediate agent with a deformable material that prevents intense grabbing in terms of adhesive wear;
- anti-friction solid lubricant, minimizing the pickup of a deformable material on the tool in the stamping process;
- barrier layer that prevents the temper of the surface layer of the tool as a result of exposure of elevated thermal loads;
- barrier layer, that is able to inhibit or stop the spread of the grid of fatigue cracks, leading to macrochipping of the working edges.

3 Technology of application of vacuum-plasma coatings on the shearing dies

As the work items of the shearing dies depending on seriation and other conditions of production organization are made of hot-short (220-250 °C) steals of the types C105W1, 105WCr6 (DIN) and semi-heatproof (490-510 °C) tool steels of the types X165CrMoV12 and K190PM (DIN), applying the traditional techniques of high-temperature vacuum-plasma coating is impractical. Even a slight rise of the temperature of coating relative to the level of acceptable heat resistance of the tool material, inevitably leads to tempering of working edges and loss of operability of the tool. At the same time, insufficient cleaning and activation of the surface often leads to delamination of the coatings in the first minutes of tool work, which is a consequence of low strength of the adhesive bond of the coating with the surfaces of the tool.

To solve the problem of temperature reduction in the process of applying of vacuum plasma coatings on die steels, was used an innovative device developed in MSTU "STANKIN", that generates a beam of accelerated neutral molecules of different gases (e.g. argon) that provide the desired heating and cleaning of the surface at the expense of its kinetic energy [6]. In this case, even with a significant decline in the value of the negative potential or even without it, it's possible to provide the necessary cleaning and activation of the surface. The heating temperature of the surface of the samples does not exceed 180 °C, that ensures the ability of continuous etching of the surface of the substrate without tempering or warpage of the working edge of the tool.

The plasma, consisting of neutral molecules of different gases, uniformly distributes over the surface of a tool of any geometrical shape and provides uniform cleaning within the line of sight of all complex-shaped surfaces.

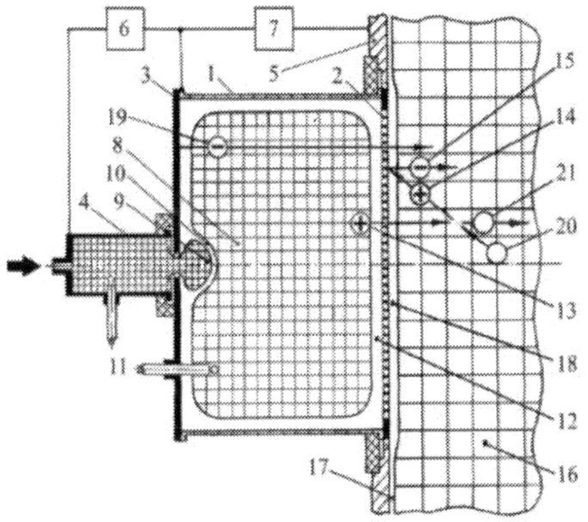

Figure 1. The design of the source of fast neutral molecules beam.

Fig. 1 demonstrates a scheme of the device. It comprises a cylindrical hollow cathode 1 with a diameter of 12 cm and length 8 cm, covered on one side with a grid 2 in the form of a disc 2 mm thick with 1224 holes with a diameter of 2 mm, and on the other side with the disk 3 with a central hole with a diameter of 8 mm.

Figure 2. The experimental unit for applying wear-resistant coatings on the tool.

Through this hole the cathode cavity communicates with a hollow anode 4 with a diameter of 2 cm and a length of 4 mm. Hollow cathode 1 is installed in the center of the end flange 5 of the vacuum chamber with a diameter of 26 cm and a length of 30 cm, pumped by a diffusion pump with liquid nitrogen trap. The gas is fed into the hollow anode 4, through a hole with a diameter of 8 mm, it flows into the hollow cathode 1 and then through the cathode grid 2 is fed into the vacuum chamber. At a sufficient flow rate of the gas the pressure inside the hollow anode 4 exceeds the pressure in hollow cathode 1 and the camera by a decade. Cathode 1, anode 4 and the chamber are made of non-magnetic stainless steel and are insulated from each other.

The discharge between the anode and the cathode is supported by the rectifier 6 with a voltage of 1.5 kV and the source 7 allows to change the potential of the cathode φ relative to the camera from 1 kV to + 1 kV. Two solenoids, covering the end flanges of the camera, create axial magnetic field with intensity up to 500 Oe, homogeneous throughout the volume of the hollow cathode and chamber, and an optical window in the center of the opposite flange of the camera allows to see the glow of the discharge inside the cathode through the grid.

At the ignition of the discharge, the cathode 1 is filled with sufficiently uniform glow of plasma 8, and in its central hole there is a brighter glow of the anode plasma 9 of double electrostatic layer 10 that penetrates into the cathode 1 from the hollow anode 4.

A part of ions 13 accelerated in the layer 12 penetrate into the chamber through the holes of the grid 2, and this leads to an increase of the potential of the space inside it. As a result of interaction of ions 13 and 14 with the chamber walls and the grid 2 appear secondary electrons 15. They are captured in an electrostatic trap of the space charge of the ions and compensate this charge, making multiple oscillations in the trap. Occurs the synthesis of the quasi-neutral plasma 16 separated from the chamber wall 5 by a layer of positive space charge 17.

On the base of the study of properties of existing promising thin-film coatings and experience in their use, were offered four different variants of wear-resistant coatings that enhance the operability of the working elements of the shear dies and the quality of fabricated parts for various character of existing loads and wear mechanisms:

1) Ti/(Ti,Zr)N coating - enhances the operability of the tool in the simultaneous occurrence of adhesive and abrasive wear. The outer layer of nitride (Ti,Zr)N increases the microhardness of the working surfaces of the tool (up to 28 hPA) and has a high resistance to molecular grabbing with the majority of processed materials. The introduction of a sublayer of titanium enhances the strength of the adhesive bond between the outer layer and the instrumental material by reducing the difference of their physical, mechanical and thermal properties, leading to the formation of large residual stresses in the coating.

2) Cr/CrN coating - enhances the performance of the tool in terms of adhesion wear and intense pickup of workpiece material on the working surface of the tool. The outer layer based on nitride CrN simultaneously increases anti-friction properties (decreasing coefficient of friction in 2 times) and increases the microhardness of the working surfaces of the tool up to 14 hPA.

3) TiN/TiCN/(Ti,Cr)N coating - enhances the operability of the tool in terms of brittle fracture (chipping) of working edges and abrasive wear. This coating

effectively inhibits the development of cracks as it consists of three alternating layers of variable hardness, significantly increases the cracking resistance and fracture toughness of the composition "coating-tool". The outer coating layer consisting of a nitride (Ti,Cr)N, increases the microhardness of the working surfaces of the tool up to 24 hPA.

4) Ti/(Ti,Al)N coating - enhances the operability of the tool in terms of frequent overheating of cutting edges and/or intensive abrasive wear. A distinctive feature of nitride (Ti,Al)N in addition to its increased microhardness (up to 37 hPA) is a high temperature resistance and the formation on the surface of a dense amorphous Al2O3 layer during heating, that serves as a kind of thermal barrier that protects the surface layers of the tool from the effects of thermal stress, which often occurs during the pressing of corrosion resistant steels and other hard materials.

4 Research of operability of the shearing dies with various vacuum-plasma coatings

The experiments simulated the different wear mechanisms, typical for the real operating conditions of the shearing dies. For this purpose were made four groups of punches and dies of different shapes and sizes made of steel X165CrMoV12, which in accordance with the developed technological principles were coated with vacuum-plasma coatings of different composition.

The tests of punching tools with coatings applied at different technological modes were carried out at cutting and punching of products of various sizes:

- coating Ti/(Ti,Zr)N - when punching holes in low-carbon steel DC04G1 (DIN) 4 mm thick in the simultaneous occurrence of adhesive and abrasive wear;
- coating Cr/CrN – when cutting the outer contour of parts ("leverage" type) in aluminum alloy G-AlMg5 (DIN) 4 mm thick in the conditions of adhesion of intensive wear and pickup of the material of the workpiece on the working surface of the tool;
- coating in TiN/TiCN/(Ti,Cr)N - when punching the outer contour of parts ("star" type) in high-carbon steel C 80W2 (DIN) 4 mm thick in terms of heavy chipping and abrasive wear of the working surfaces of the tool;
- coating in Ti/(Ti,Al)N - when cutting holes in corrosion-resistant steel X10CrNiTi18-9 (DIN) 4 mm thick in terms of frequent overheating and, as a consequence, tempering of working edges.

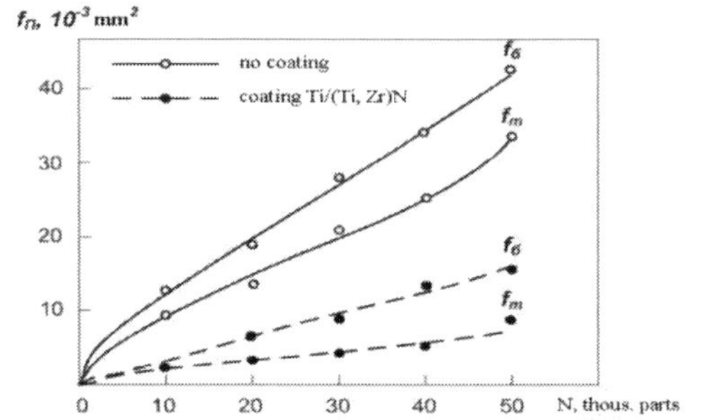

Figure 3. The dependence of the butt f_m and side $f_б$ wear of the punch when punching holes in the material DC04G1 on the number of stamped details.

Fig. 3 shows fragments of experimental studies of dependence of the wear of the punch with different coatings on the number of stamped parts made of steel DC04G1. The results of experiments conducted on the samples of aluminum alloy G-AlMg5, which has the most intense burr-formation, also showed a significant effect from the use of proposed coatings.

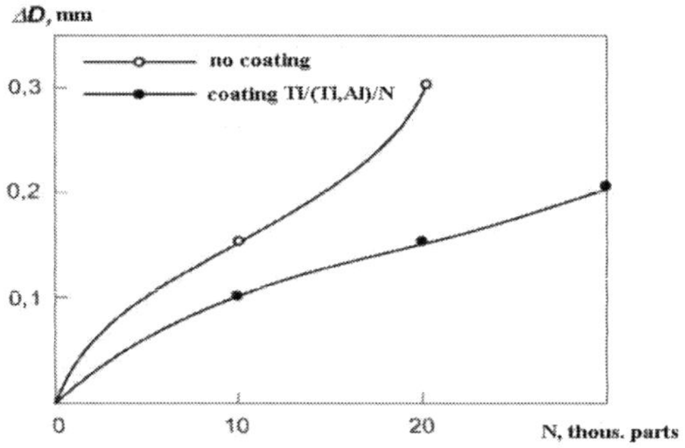

Figure 4. The dependence of size of the holes in the material X10CrNiTi18-9 on the number.

Fig. 4 shows the change of the size of the hole, punched in steel X10CrNiTi18-9, depending on the number of stamped parts. It is noticeable that at the time of punching of 20 thousand holes, the application of the proposed coating allows to halve the deviation of the size of the hole.

5 Conclusion

To ensure the ability of the application of high quality coatings on hot-short steels and semi-heatproof shear die steels can be used the proposed technological principles of low-temperature deposition of multilayer vacuum-plasma coatings in the mode of assisting with the beam of accelerated molecules of argon. It is proposed to use four typical variants of architecture of vacuum-plasma coatings - Ti/(Ti,Zr)N, Cr/CrN, TiN/TiCN/(Ti,Cr)N and Ti/(Ti,Al)N.

On the basis of processing of the results of operational tests was established the fact that the application of a vacuum plasma coating under optimal conditions reduces the wear rate of the shearing dies more than in 2.5 times;

DOI: 10.1051/matecconf/20166503008

the height of the burr formed on the end butt of the surface of the parts reduces in 3 times, and the deviation of the sizes of the parts in 2 times.

The work is performed with financial support of the Russian Science Foundation within the Agreement No. 14-29-00297 of August 6, 2014.

The work is carried out on the equipment of the Center of collective use of MSTU "STANKIN" with financial support of the Ministry of Education and Science of Russian Federation, the Agreement No. 14.593.21.0004 of 04.12.2014, the unique identifier of the project RFMEFI59314X0004.

References

1. S.N. Grigoriev, M.A. Volosova, Complex surface hardening of oxide-carbide ceramic cutting tools, Rus. Eng. Res. 9 (2005) 7-12.

2. A.A. Andreev, I.V. Serdyuk, V.M. Beresnev, M.A. Volosova, S.N. Grigoriev, D.A. Kolesnikov, A.D. Pogrebnyak, O.V. Sobol, P.V Turbin, Tribological characteristics of nanocomposite vacuum-plasma TI-HF, TI-HF-N, and TI-HF-SI-N coatings, J. of Frict. and Wear. Vol. 34. 3 (2013) 175-182.

3. A.S. Vereschaka, S.N. Grigoriev, E.S. Sotova, A.A. Vereschaka, V.P. Tabakov, M.Yu. Kulikov, Improving the efficiency of the cutting tool made of ceramic when machining hardened steel by applying nano-dispersed multi-layered coatings. Key Eng. Mat. Vol. 581 (2014) 68-73.

4. S.N. Grigoriev, Yu.A. Melnik, A.C. Metel, Broad fast neutral molecule beam sources for industrial-scale beam-assisted deposition, Surf. and Coat. Tech. Vol. 156 №1-3 (2002) 44-49.

5. A.S. Vereschaka, S.N. Grigoriev, E.S. Sotova, A.A. Vereschaka, Improving the efficiency of the cutting tools made of mixed ceramics by applying modifying nano-scale multilayered coatings, Adv. Mat. Res. Vol. 712-715 (2013) 391-394.

6. A.C. Metel, V.P. Bolbukov, M.A. Volosova, S.N. Grigoriev, Yu.A. Melnik, Source of metal atoms and fast gas molecules for coating deposition on complex shaped dielectric products, Surf. and Coat. Tech. Vol. 225 (2013) 34-39.

Research on the Prediction Model of CPU Utilization Based on ARIMA-BP Neural Network

Jina Wang[1], Yongming Yan[2,a] and Jun Guo[2]

[1] *Liaoning Software Testing Center, Shenyang, China*
[2] *College of Information Science & Technology Engineering, Northeastern University, Shenyang, China*

Abstract. The dynamic deployment technology of the virtual machine is one of the current cloud computing research focuses. The traditional methods mainly work after the degradation of the service performance that usually lag. To solve the problem a new prediction model based on the CPU utilization is constructed in this paper. A reference offered by the new prediction model of the CPU utilization is provided to the VM dynamic deployment process which will speed to finish the deployment process before the degradation of the service performance. By this method it not only ensure the quality of services but also improve the server performance and resource utilization. The new prediction method of the CPU utilization based on the ARIMA-BP neural network mainly include four parts: preprocess the collected data, build the predictive model of ARIMA-BP neural network, modify the nonlinear residuals of the time series by the BP prediction algorithm and obtain the prediction results by analyzing the above data comprehensively.

1 Introduction

In the recent cloud computing research field, the virtual resource management problem has become a hotspot. The VM dynamic deployment technique is one of the key virtual resource management technique. Many experts have done a lot of research about it.

At present, the study of the VM resource dynamic deployment focus on the fine-grained resource adjustment strategy. The current workload data and the future change trend data are needed if a predictive and accurate resource adjustment plan is to be made. Daniel A Menasce[1] put forward a dynamic resource adjustment plan following the changing workload by the CPU priority or the CPU shares. ZHAO Weiming[2] presented a new theory to improve memory utilization by predicting memory usage of each VMs. Anton Beloglazov[3] proposed a set of the heuristic algorithms to select the objective VM for migration when the VM violate the SLA. A new distributed management strategy select the VM whose CPU usage rate is the maximum in the abnormal server in the reference 4. If after migrating the CPU usage rate still be high, it would continue to migrate the VM whose CPU usage rate is the maximum in the rest VMs until recovering the server. Compared with the reference 4, some researchers put forward a new algorithm to select the VM for migration by predicting the workload trend so that it can avoid the instantaneous peak load [5].

[a] Corresponding author: yym_sy@163.com

2 The Research Method of the CPU Utilization Prediction Process

With the ongoing changes of the workload, the service performance should be ensured by adjusting the resource of the VM. After adding the predictive method, the VM dynamic deployment plan can avoid the performance fluctuation caused by the lag of the optimization process. By the historical data a predictive model of the future CPU utilization can be set up. With the predictive information, more supports can be received by the dynamic deployment plan.

So far, the method of the weighted average to predicted single resource change trend is widely used in the existing models, but the method using the composite model to predict the CPU utilization by the time series analysis technique is adopted in few researches. The research emphasis in the former is to design the weighting efficient system. A good weighting efficient system means an accurate predicted result. But with much subjectivity and arbitrariness it is very difficult to design a good system. By researching, it is found that a time series process consists of a linear structure and a nonlinear structure that means the change trend rule can be found easily by the time series.

The linear prediction model is mainly adopted in the time series prediction method so that the method cannot process the nonlinear data well. Compared with the traditional data processing algorithms, BP neural network is an effective nonlinear modeling method. It has great advantages in processing the feature-undefined data that with much randomness, nonlinearity and noisy. Implied non-linear relationships can be excavated out from the database by the BP model. But it is weaker in process the linear data than the traditional methods such as time series model. In fact most server CPU utilization sequences usually consist of linear time sequence part and nonlinear time sequence part that means it mix the character of linear and nonlinear. So the composite model combining with BP neural network and time series prediction method can improve the predictive result.

According to the analysis, a new predictive model combing with ARIMA model and BP neural network to predict the future CPU utilization is put forward in this paper.

The CPU utilization is fitted by the ARIMA prediction technique in the new model, then the nonlinear residual is calculated by the BP neural network and finally the real predictive result is obtained by adding the both parts. The relevant technical theories are introduced as follows.

2.1 BP Neural Network

BP(Back propagation), an abbreviation for "backward propagation of errors", is a common method of training artificial neural networks used in conjunction with an optimization method such as gradient descent.

Each propagation involves the following steps:

(1) Forward propagation of a training pattern's input through the neural network in order to generate the propagation's output activations.
(2) Backward propagation of the propagation's output activations through the neural network using the training pattern target in order to generate the deltas of all output and hidden neurons.

For each weight-synapse follow the following steps:

(1) Multiply its output delta and input activation to get the gradient of the weight.
(2) Subtract a ratio (percentage) of the gradient from the weight.

2.2 Time Series Prediction

The steps of building the ARIMA(p,d,q) model are shown as follows:

(1) Collect the time series data set;
(2) Draw out the sequence diagrams, autocorrelation function diagram and partial autocorrelation function diagrams of the observed data. And judge the series whether is stationary series or not. If not it should be process by the differential computing process to get the stationary series.
(3) Calculate the Q_{LB} statistics of the observed series, and verify the stochastic feature of the series.
(4) The ARIMA(p,d,q) model is built by combine the feature values of the AR function and the partial AR function. Select a proper order value and then estimate model parameters.
(5) Calculate the Q_{LB} statistics of the residual sequence. Only when the residual sequence is the white noise sequence the built-model is a effective model.
(6) By the forecasting model the prediction values that is the important data for us to make decisions can be obtained.

3 The Modeling Method of the ARIMA-BP Model

3.1 Building the ARIMA(p, d, q) Model.

Step1: Collect the server CPU utilization data. And mark them as original series after preprocess them.

$$y(t) = \{y_1, y_2, \ldots, y_n\} \quad (1)$$

Step2: The original series should be smoothened. In this paper the autocorrelation is the index to measure the degree of tranquilization.

The autocorrelation is the ratio of the covariance of the h-lag time series to the total variance of the time series: $(\mathrm{ACs})\ \rho_h = \gamma_h / \sigma^2$. Where the γ_h can be obtain by follow formula.

$$\gamma_h = \frac{1}{T}\sum_{t=h+1}^{T}(y_t - \bar{y})(y_{t-h} - \bar{y}) \quad (2)$$

Where the $\bar{y} = T^{-1}\sum_{t=1}^{T} y_t$ is the sample meas. The autocorrelation descript the relation between the time series values and its h-lag period. One of the most important characteristic for the stable sequence is that the autocorrelation function ρ_h is rapidly descend to 0 by the increase of the h-lag period. For the unstable time series the union-change will be week.

If the time series can be proved to be unstable, its unstable features can be separate out so that the stable time series can be obtained. In this paper the method that combining the logarithm transformation and the difference algebra is used to smoothen the time series.

Step3: Calculate the autocorrelation coefficients and the part autocorrelation coefficients. And then build the time series model by the identification rule. After calculating, the autocorrelation coefficients and the part autocorrelation coefficients should follow the ARMA(p,q). Finally the order parameters will be make sure by the AIC rules. In this paper we calculate the order values are p=1, q=1. Because the time series is just with the first difference, the final model form is ARIMA (1, 1, 1)

Step4 : It is very significant to estimate the parameters. After getting the stable time series, the follow method will be used to build the ARMA(p, q) model:

$$u_t = y_t - \varphi y_{t-1} - \cdots - \varphi_p y_{t-p} + \theta_1 u_{t-1} + \cdots + \theta_q u_{t-q} \quad (3)$$

Work out the $u_t = u_t(\varphi, \theta)$. Where the t=p+1,...,T. The purpose of the parameter estimation is to get the min sum square by finding a set of parameters $\left(\varphi_1, \varphi_2, \ldots, \varphi_p, \theta_1, \theta_2, \ldots, \theta_q\right)$

$$S(\varphi, \theta) = \sum_{t=p+1}^{n} u_t^2 (\varphi, \theta) \tag{4}$$

Step5: And then the model need to be checked. Calculate the residual sequence. The model can be proved to be accuracy only when the residual sequence is white sequence.

Step6: Predicting the CPU future utilization by the change rules finding by the fitting model. The predicted data are the key reference for the future decisions.

3.2 Modifying the Residual Error Correction by the BP Neural Network

Step1: Calculate the residual sequence of the CPU utilization time series.

$$N_i = y_i - L'_{i'}, \left(i = 1, 2, 3, \ldots, n\right) \tag{5}$$

Where the $\{L'_1, L'_2, \ldots, L'_n\}$ is the predicted server CPU utilization data getting by the ARIMA(p, d, q) model. Where the $\{y_1, y_2, \ldots, y_n\}$ is the real server CPU utilization data.

Step2: Normalize the residual sequence so that is can be process in the sigmoid function.

$$N_i = \left(N_i - s\right) / \left(t - s\right) \tag{6}$$

Where the $s=(9(N_i)_{min}-(Ni)_{max})/8$ and $t=(9(N_i)_{max}-(Ni)_{min})/8$. Where the $(Ni)_{max}$ is the maximum value of the residual sequence and where the $(Ni)_{min}$ is the minimum value of the residual sequence.

Step3: Define the structure of the BP neural network

Analyzing the residual sequence of the server CPU utilization is the main function of the neural network. In this paper the input structure is single-input that means the number of the input layer neuron is 1 and the same with the number of the output layer neuron. The number of the hidden neurons is define as follow formula:

$$I = \sqrt{\left(m + n\right)} + a \tag{7}$$

Where the m is the number of the input layer. Where the n is the number of the output layer. Where the a is a random constant range from 0 to 10.

Step4: Define the trained parameters of the BP neural network: The main parameters of BP neural network are the transfer function of the hidden layer, the transfer function of the output layer, the learning rate and the iterations.

Step5: The neural network need to be trained several times. The different network need to be compared so that the best one can be selected out. And at last the residual sequence N'_t can be obtained by the network simulation.

3.3 The prediction of the server CPU utilization

The final predicted data of the server CPU utilization y'_t that based on the ARIMA-BP neural network algorithm can be obtained by adding the initial predicted data of the server CPU utilization y'_t predicted by the ARIMA(p,d,q) and the residual sequence N'_t predicted by the BP neural network.

4 Analysis of Experimental Results

4.1 The experimental steps

(1) Filter the Data.

After be preprocessed the data of the CPU utilization monitored by the monitoring program is the original time series of the server CPU utilization. The interval is 15 seconds. Select 300 records with the interval to predict the CPU utilization in future 30 intervals. The time series of the server CPU utilization is shown in Figure.1.

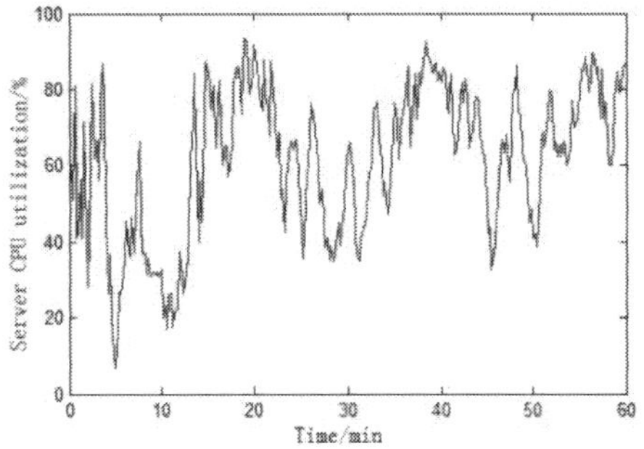

Figure 1. CPU utilization of physical server.

(2) Smoothen the original time series.

In this paper the method that combining the logarithm transformation and the difference algebra is used to smoothen the time series. The processed result is shown in Figure. 2.

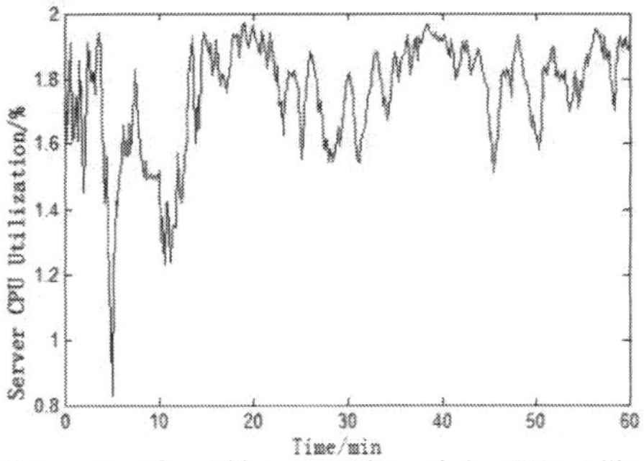

Figure 2. After taking logarithm of the CPU utilization of physical server.

(3) Built the ARIMA(p, d, q) model.

The method of building the model is combined the contents in the chapter 4.4.2 and the AIC rules. It is obviously when R=1 and M=1, the values of AIC and BIC can be got the minimum. So the model ARIMA(1,1,1) is chosen in the experiment. After be checking the residual sequence of the model can be proved to be white noise sequence that means the model is effective.

R=0,M=0,AIC=2202.480317,BIC=2209.881204
R=0,M=1,AIC=2202.148436,BIC=2213.249767

R=0,M=2,AIC=2200.471689,BIC=2215.273463
R=0,M=3,AIC=2200.087553,BIC=2218.589771
R=1,M=0,AIC=2202.601972,BIC=2213.703303

(4) Modify the residual sequence by the BP neural network

The residual sequence of the predicted value need to be worked out t and analyzed. And then the residual sequence need to be modified. The final predicted data of the server CPU utilization can be obtained by adding the initial predicted data of the server CPU utilization predicted by the ARIMA(1,1,1) and the analyzed result of the residual sequence.

4.2 The analysis of experiment results

The details of the IBM server is shown as follows: eight-core(two Intel Xeon E5506 2.13GHz), 16 GB 2-channel memory (1333MHz), 500G hard disk and with Linux system. For to creating and managing five virtual machines, the Xen system is installed in the server. And the Xen system is also used to manage the resource of the server. The resource of the Xen's virtual domains should be allocated when starting the virtual domains. The physical resource of the server will be shared by all the virtual domains in the form of partitions. All the utilization of the server resource will be monitored by the Xen system.

The composite method ARIMA(p,d,q)-BP is compared to the signal method ARIMA(p,d,q) and the signal method BP neural network. From the following data figure and tables it can be seen that the result that come from the composite model is more close to the real value.

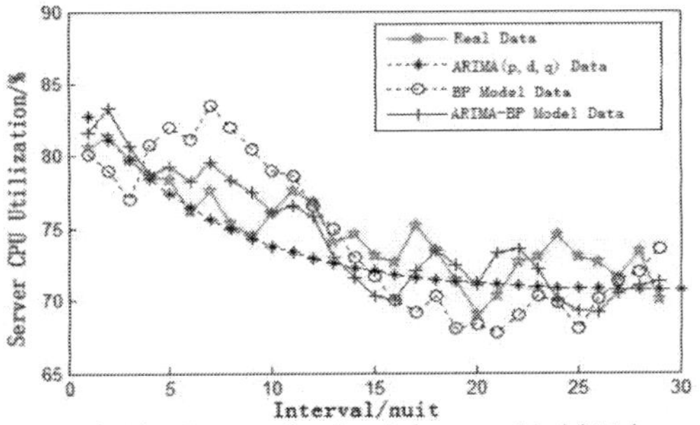

Figure 3. The Comparative Result between Model Value and Real Value of Three Predicted.

Table 1. The Errors Comparative Result of Three Predicted Algorithm.

Algorithm	mean relative error	mean square error	mean absolute error
ARIMA	0.0370	0.303	0.208
BP	0.0265	0.228	0.125
ARIMA -BP	0.0186	0.126	0.097

After Analyzing the data in table 1, it can be seen that the ARIMA(p,d,q)-BP model is more accurate than the single model. The mean relative error separately decreased from 0.0370 and 0.0265 down to 0.0186. The mean square error separately decreased from 0.303 and 0.0228 down to 0.0126. The mean absolute error separately decreased from 0.208 and 0.125 down to 0.097. So the composite model is proved to be more accurate.

5 Conclusion

In this paper, the new composite model combining with the ARIMA(p,d,q) model and the BP neural network is proposed. With the composite model, the future change trend rules of the server CPU utilization can be found. The prediction value of the future server CPU utilization predicted by the new model can be more close to the real value. With the accurate predictive data the strategy of the VM resource dynamic deployment will be more useful.

Acknowledgments

This work was supported by the National Natural Science Foundation Program of China (61572117, 61300019, 61370155) and the Special Fund for Fundamental Research of Central Universities of Northeastern University (N120804001, N120204003, N140406002).

References

1. Menasce D A, Bennani M N. Autonomic virtualized environments[C].Autonomic and Autonomous Systems, 2006. ICAS'06. 2006 International Conference on. IEEE, 2006: 28-28.
2. Zhao W, Wang Z, Luo Y. Dynamic memory balancing for virtual machines[J]. ACM SIGOPS Operating Systems Review, 2009, 43(3): 37-47.
3. Beloglazov A, Abawajy J, Buyya R. Energy-aware resource allocation heuristics for efficient management of data centers for cloud computing[J]. Future Generation Computer Systems 2011; doi:10.1016/j.future.2011.04.017.
4. MA Fei, LIU Feng, LIU Zhen Efficiency Energy-saving Distributed Management Method of Virtual Machine in Cloud Computing [J]. Computer Engineering, 2012, 38(11).
5. HU Zhi-gang, OUYANG Cheng, YAN Chao-kun Resource Load Balancing Method for Energy-consumption Reducing in Cloud Environment[J] Computer Engineering, 38(5): 53-55.

Reduction of temperature in silicon photovoltaic module using thermal radiation coating

Satoshi Nakamura [1] and Kensuke Nishioka [1,a]

[1] *Faculty of Engineering, University of Miyazaki, 1-1 Gakuen Kibanadai Nishi Miyazaki 889-2192, Japan*

Abstract. The temperature of solar cells increases under the actual operating conditions, and the conversion efficiency of solar cells decreases with increasing temperature. In this study, a thermal radiation layer was coated on the back sheet of the PV module by a spray coating method and the effect was evaluated. The thickness of the layer was 30 μm. The temperature of the PV module with the thermal radiation coating was lower than that of the PV module without the thermal radiation coating. And the operating temperature range of the PV module with the thermal radiation coating was decreased by 2~3°C. The open-circuit voltage of the PV module with the thermal radiation coating was 0.1 V higher than that of the module without the coating due to the thermal radiation coating.

1 Introduction

Photovoltaic (PV) power generation is becoming widespread as a clean and gentle energy source for the earth. Though the performance of solar cell is mainly evaluated under the standard test condition (STC: 1kW/m^2 irradiation, 25°C module temperature, and AM1.5 global spectrum), operation under various environments is required for PV systems, and environmental factors such as solar irradiation and module temperature seriously influence the generating performance of the systems

The temperature of solar cells increases under the actual operating conditions, and the conversion efficiency of solar cells decreases with increasing temperature [1-11]. It is therefore very important to reduce the cell temperature in PV modules.

In this study, a thermal radiation layer was coated on the back sheet of a PV module and the effect was evaluated.

2 Experimental procedure

Figure 1 shows the single crystalline silicon PV modules (area: 1200 mm × 527 mm) evaluated in this study. The output characteristics of PV modules with and without thermal radiation coating were compared. A thermal radiation layer (Pelcool (R), PELNOX Ltd.) was coated on the back sheet of the PV module by a spray coating method. The thickness of the layer was 30 μm. The thermal radiation layer consisted of acrylate resin and inorganic fillers. The fillers were selected to radiate the heat, particularly in the temperature range from 40 to 100°C, which is the main range of operating temperature for the PV module. The thermal emissivity of the layer is 0.95. In order to detect the temperature of the PV module (T_{mod}), temperature sensors (Pt100) were set on the back surface of the PV modules. The current-voltage (I-V) characteristics were measured using an I-V curve tracer (MP-160, EKO). The modules with and without the thermal radiation coating were evaluated at University of Miyazaki (Miyazaki, Japan).

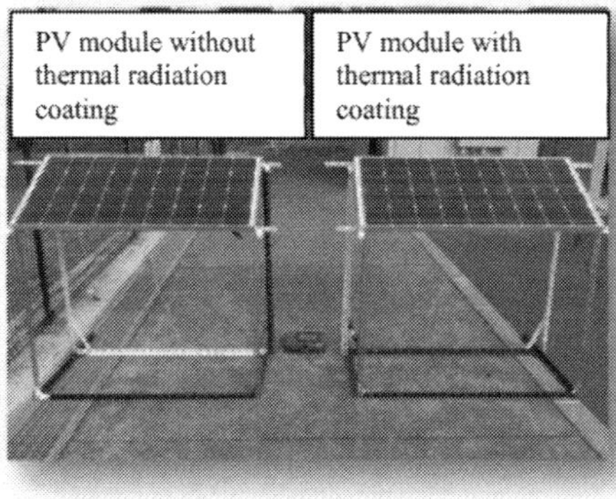

Figure 1. Single crystalline silicon PV modules (area: 1200 mm × 527 mm) evaluated in this study.

3 Results and discussion

Figure 2 shows the temperatures of the PV modules (T_{mod}) with and without thermal radiation coating. Figure 2 also shows the ambient temperature. The ambient temperature was stable during the measurement period. The temperature of the PV module with the thermal radiation coating was lower than that of the PV module

[a] nishioka@cc.miyazaki-u.ac.jp

DOI: 10.1051/matecconf/20166504001

without the thermal radiation coating. The effect of the high-radiation layer was remarkable.

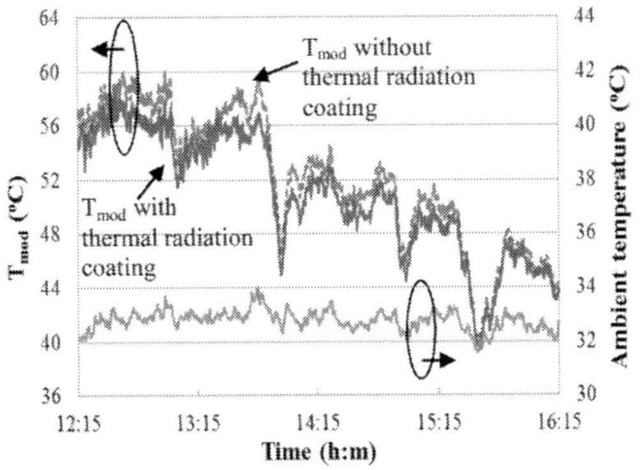

Figure 2. Temperatures of the PV modules with and without thermal radiation coating.

Figure 3 shows the T_{mod} of the PV module without the thermal radiation coating and the difference in temperature between T_{mod} of the PV modules with and without thermal radiation coating (T_{mod} without coating – T_{mod} with coating). The maximum difference of 3.38°C was observed at 12:57. It is found that the difference in temperature between T_{mod} of the PV modules with and without thermal radiation coating increases with increasing T_{mod} of the PV module without thermal radiation coating.

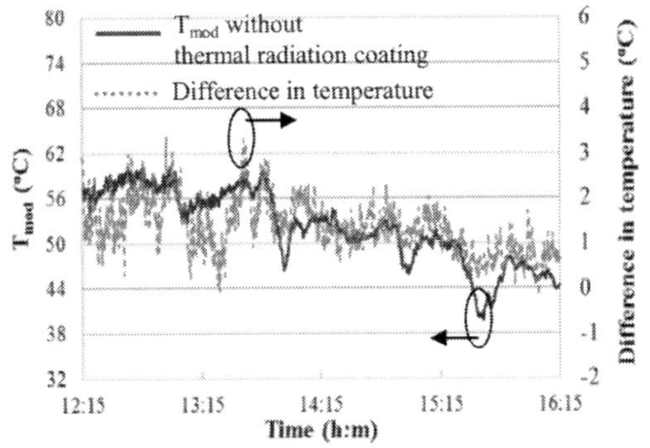

Figure 3. T_{mod} of the PV module without thermal radiation coating and the difference in temperature between T_{mod} of the PV modules with and without thermal radiation coating.

When we define a backside of PV module [area: A_1 (m^2), emissivity: ε_1] and an environment [area: infinite], the radiation heat from the backside of PV module to the environment Q_{12} (W) is given by

$$Q_{12} = A_1 \varepsilon_1 \sigma \left(T_1^4 - T_2^4\right) \quad (1)$$

where σ, T_1, and T_2 are the Stefan–Boltzmann constant (5.67 × 10-8 W/m^2·K^4), the absolute temperature (K) of the PV module, and the absolute temperature (K) of the environment, respectively. It was found that the radiation heat increased with increasing emissivity, and the high emissivity of the thermal radiation layer enhanced the heat radiation from the module to the environment. Q_{12} (W) is higher when T_1 is higher. Therefore, the difference in temperature between T_{mod} of the PV modules with and without thermal radiation coating increased with increasing T_{mod} of the PV module without thermal radiation coating.

Figure 4 shows the open-circuit voltage (V_{oc}) of the PV modules with and without the thermal radiation coating. V_{oc} of the PV module with the thermal radiation coating was 0.1 V higher than that of the module without the coating during the test period. The V_{oc} of PV modules decreases with increasing temperature.

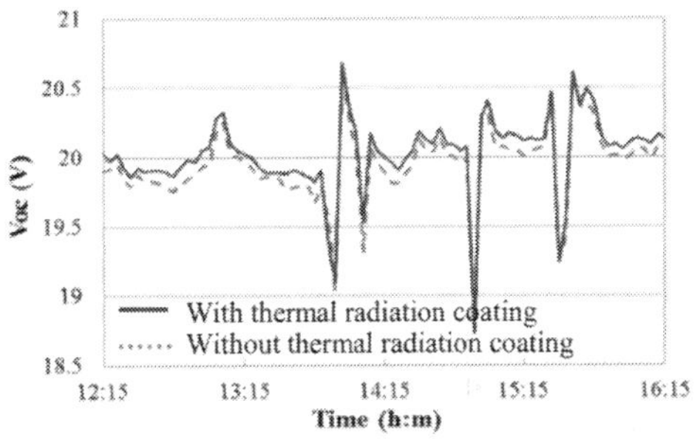

Figure 4. Open-circuit voltage (V_{oc}) of the PV modules with and without the thermal radiation coating.

The I-V characteristics of the solar cell are expressed by

$$I = I_0 \left\{ \exp\left(\frac{qV}{n_D kT} \right) - 1 \right\} - I_{sc}, \quad (2)$$

where I_{sc}, I_0, q, n_D, k, and T are the short-circuit current, saturation current, elementary charge, diode ideality factor, Boltzmann constant, and absolute temperature, respectively [12].

From Eq. (2), V_{oc} ($I = 0$) is given by

$$V_{oc} = \frac{n_D kT}{q} \ln\left(\frac{I_{sc}}{I_0} + 1 \right). \quad (3)$$

From Eq. (3), the temperature characteristic of saturation current (I_0) markedly influences the temperature characteristic of V_{oc}. The saturation current density (J_0) is given by

$$J_0 = qn_i^2\left(\frac{D_e}{N_A W_p} + \frac{D_h}{N_D W_n}\right) \quad (4)$$

where n_i is the intrinsic carrier concentration, N_A and N_D are the acceptor and donor concentrations, respectively, W_p and W_n are the thicknesses of the *p* and *n* neutral regions, respectively, and D_e and D_h are the diffusion constants of electrons and holes, respectively [13]. J_0 strongly depends on *T* through its proportionality to the square of n_i, which is expressed by

$$n_i^2 = 4M_c M_v (2\pi kT/h^2)^3 (m_e^* m_h^*)^{3/2} \exp(-E_g/kT), \quad (5)$$

where M_c and M_v are the number of equivalent minima in the conduction and valence bands, respectively, *h* is Planck's constant, and m_e* and m_h* are the effective masses of electrons and holes, respectively [14].

From Eqs. (3)–(5), it is found that the decrease in V_{oc} with increasing temperature arises mainly from the change in n_i. The value of J_0 increases exponentially with decreasing 1/T, and V_{oc} decreases linearly with increasing T.

V_{oc} of the PV module with the thermal radiation coating was higher due to the cooling effect of the thermal radiation coating.

Figure 5 shows the conversion efficiency of the PV modules with and without the thermal radiation coating. The conversion efficiency decreased linearly with increasing temperature. The data for the PV module with the thermal radiation coating existed in the low temperature range owing to the heat-release effect of the coating. As shown in Fig. 3, the cell temperature of the PV module with the thermal radiation coating was 2~3°C lower than that of the module without thermal radiation coating. Eventually, the temperature range of the PV module with the thermal radiation coating was decreased by 2~3°C.

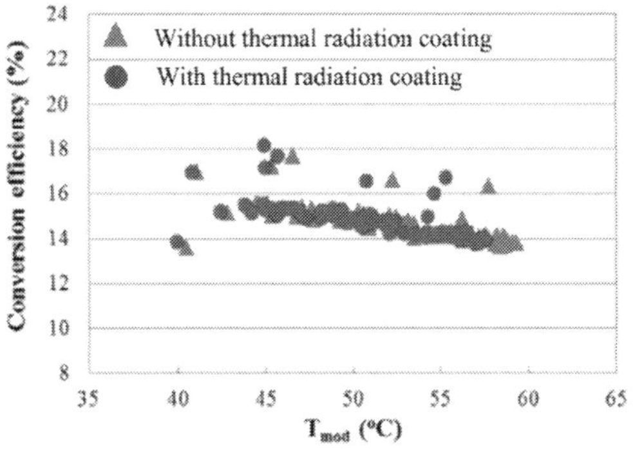

Figure 5. Conversion efficiency of the PV modules with and without the thermal radiation coating.

A high-efficiency PV module can be achieved with a combination of cell and module technologies. In this study, a new simple coating technology for handling heat radiation was developed. By adopting the thermal radiation coating for the PV module fabrication, the module efficiency was easily improved.

4 Summary

Single crystalline silicon PV modules (area: 1200 mm × 527 mm) were prepared and evaluated. A thermal radiation layer (Pelcool (R), PELNOX Ltd.) was coated on the back sheet of the PV module. The thickness of the layer was 30 μm. The thermal radiation layer consisted of acrylate resin and inorganic fillers. The fillers were selected to radiate the heat, particularly in the temperature range from 40 to 100°C, which is the main range of operating temperature for the PV module. The thermal emissivity of the layer is 0.95. In order to detect the temperature of the PV module (T_{mod}), temperature sensors (Pt100) were set on the back surface of the PV modules. The temperature of the PV module with the thermal radiation coating was lower than that of the PV module without the thermal radiation coating. V_{oc} of the PV module with the thermal radiation coating was 0.1 V higher than that of the module without the coating during the test period. The operating temperature range of the PV module with the thermal radiation coating was decreased by 2~3°C. By adopting the thermal radiation coating, the module efficiency was easily improved. A new simple coating technology for handling heat radiation was developed. By adopting the thermal radiation coating for the PV module fabrication, the module efficiency was easily improved.

References

1. K. Nishioka, T. Hatayama, Y. Uraoka, T. Fuyuki, R. Hagihara, M. Watanabe, Solar Energy Materials and Solar Cells **75**, 665 (2003)
2. K. Nishioka, N. Sakitani, K. Kurobe, Y. Yamamoto, Y. Ishikawa, Y. Uraoka, T. Fuyuki, Jpn. J. Appl. Phys. **42**, 7175 (2003)
3. K. Nishioka, T. Takamoto, T. Agui, M. Kaneiwa, Y. Uraoka, T. Fuyuki, Solar Energy Materials and Solar Cells **85**, 429 (2005)
4. K. Nishioka, T. Takamoto, T. Agui, M. Kaneiwa, Y. Uraoka, T. Fuyuki, Solar Energy Materials and Solar Cells **90**, 57 (2006)
5. Kensuke Nishioka, Tsuyoshi Sueto, Masaki Uchida, Yasuyuki Ota, Journal of Electronic Materials **39**, 704 (2010)
6. Kensuke Nishioka, Yasuyuki Ota, Kazuyuki Tamura, Kenji Araki, Surface and Coatings Technology **215**, 472 (2013)
7. Yasuyuki Ota, Tsuyoshi Sueto, Hirokazu Nagai, Kenji Araki, Kensuke Nishioka, Jpn. J. Appl. Phys. **52**, 04CR03 (2013)
8. K. Nomura, Y. Ota, T. Minemoto, K. Nishioka, ACTA PHYSICA POLONICA A **125**, 1021 (2014)

9. Kei Nomura, Yasuyuki Ota, Takashi Minemoto, Kensuke Nishioka, physica status solidi (c) **11**, 1427 (2014)
10. Yasuyuki Ota, Hirokazu Nagai, Kenji Araki, Kensuke Nishioka, Jpn. J. Appl. Phys. **53**, 122301 (2014)
11. Yuya Sakurada, Yasuyuki Ota, Kensuke Nishioka, Jpn. J. Appl. Phys. **50**, 04DP13 (2011)
12. S. M. Sze, *Physics of Semiconductor Devices* (Wiley-Interscience Publication, 1981)
13. P. Antognetti and G. Massobrio, *Semiconductor Device Modeling with SPICE* (McGraw-Hill, 1988)
14. A. L. Fahrenbruch and R. H. Bube, *Fundamentals of Solar Cells* (Acadmic Press, Inc., 1983)

Designing of long wavelength cut thin film filter for temperature reduction of concentrator photovoltaic

Nawwar Ahmad [1], Keisuke Hatakeyama [1], Yasuyuki Ota [1] and Kensuke Nishioka [1,a]

[1] *Faculty of Engineering, University of Miyazaki, 1-1, Gakuen Kibanadai Nishi, Miyazaki 889-2192, Japan.*

Abstract. This paper presents a comparison of three-dimensional simulation for concentrator photovoltaic module using two types of multi-junction solar cell. Each had its own range of spectral response and based on that range a thin film filter was developed for each case to reflect the unused spectral of the solar spectrum and allowed the desired spectrum to reach the solar cell. The thin film was deposited on a secondary optical element that was used to homogenize the irradiance distribution on the solar cell. A thermal simulation was conducted to compare the resulted decrease in cell temperature due to the use of the thin film for each case.

1 Introduction

Concentrator photovoltaic (CPV) system is considered one of the most promising technologies in the field of solar energy which have received an increasing attention in R&D in recent years [1]. A CPV system utilizes the inexpensive optical elements, such as Fresnel lenses to concentrate sunlight into a small-size III-V multi-junction solar cell [2]. Multi-junction solar cells are influenced by their operating temperature. Decreasing solar cell temperature can have great impact on the open circuit voltage, maximum power point and efficiency of the multi-junction cell. In this paper we propose a Fresnel lens optical concentration system with concentration ratio of 1322 times.

The optical systems concentrate solar irradiance on two types of solar cells. Each has its own spectral response range, the first cell is InGaP/InGaAs/Ge triple junction solar cell with high spectral response in the range from 400 nm to 1700 nm [3], and the other is InGaP/GaAs/InGaAs triple junction solar cell with high spectral response in the range from 400 nm to 1100 nm [4]. We used thin film filter to utilize the part of solar spectrum which is beneficial to the solar cell while reflecting the rest of the long wavelength spectrum which will result in reduce of the cell temperature.

2 Optical model

Ray-trace calculation was conducted for the optical system using commercial optical simulation software ZEMAX. The optical system consists of typical flat Fresnel lens as primary optical element with dimensions of 200 mm × 200 mm as incident ray area (focal length = 420 mm), and secondary optical element (homogenizer).

Figure 1 shows the schematic diagram of the CPV optical structure. The homogenizer (14 mm×14 mm as entry aperture area, 5.5 mm×5.5 mm as exit aperture area, and 40 mm in height) was set at the focal point of the Fresnel lens. The used thin film was deposited at the entry aperture of the homogenizer, and the resulted intensity distribution of the concentrated light was analysed. The distance between Fresnel lens and solar cell was 460 mm. the geometrical concentration ratio for this system was 1,322 times.

The spectral irradiance using ray trace simulation was AM1.5 (total power: 900 W/m^2). Ray trace was carried out in the wave length from 300 to 2500 nm to show the effect of the thin film on the overall performance of the system.

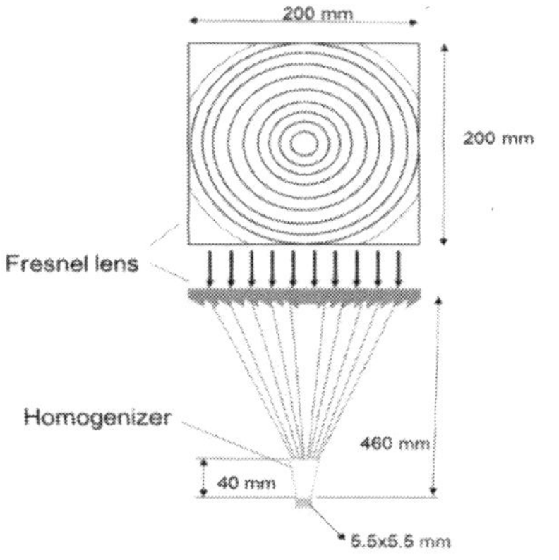

Figure 1. Schematic Diagram of the CPV optical structure.

[a] nishioka@cc.miyazaki-u.ac.jp

3 Thin film Design

For designing the thin film there is different traditional optimization method wildly accepted and used, nevertheless most used optimization technique for thin film are greatly affected by the initial structure of the used film, which may cause a local convergence, Tikhonravov et al. [5,6] has developed a powerful optimization method called Needle optimization technique. The two materials in the filter are H which has the high refractive index (in our case TiO_2), while L has the low refractive index material (in our case SiO_2).

Using needle optimizing and TFcalc (commercial thin film design software), we developed two filters; the first filter was used for the InGaP/InGaAs/Ge triple junction solar cell concentrator module and it only allowed the spectrum in the range from 400 to 1700 nm to be transmitted onto the solar cell while reflecting the rest of the long wavelength spectrum. The resulted structure consisted of 65 layers and had a thickness of 3490.4 nm. It's expressed as follow:

S 731.37L 4.15H 36.04L 5.42H 22.65L 7.46H 17.42L 9.99H 12.16L 16.17H 11.12L 111.53H 14.33L 20.22H 20.39L 10.89H 42.16L 6.5H 254.28L 6.41H 39.05L 9.16H 18.17L 20.3H 13.99L 130.37H 15.79L 25.23H 42.99L 10.24H 250.9L 8.42H 44.82L 23.89H 15.7L 137.42H 15.15L 23.01H 43.1L 8.07H 260.08L 8.52H 41.91L 24.87H 14.18L 130.73H 14L 16.6H 15.96L 7.22H 40L 4.06H 274.14L 5.24H 32.43L 7.28H 14.9L 19.55H 6.57L 135.53H 14.98L 14.43H 25.97L 11.64H 93.2L A,

where S is the substrate in this case it's BK7 and A stands for air.

As for the thin film used for the InGaP/GaAs/InGaAs triple junction solar cell it transmitted the spectrum in the range from 400 to 1100 nm and reflected the rest of the long wavelength spectrum. It's expressed as:

S 6.61H 38.65L 12.67H 22.3L 14.71H 22.05L 12.98H 29.1L 9.05H 50.39L 2.82H 452.84L 9.08H 38.66L 25.18H 12.45L 133.25H 7.69L 22.55H 30.71L 9.82H 230.37L 7.52H 49.58L 17.06H 27.69L 25.03H 18.8L 31.87H 9.84L 150.55H 6.26L 23.86H 32.29L 11.67H 215.82L 15.27H 26.04L 117.25H 15.75L 17.37H 33.9L 5.67H 488.99L 6.05H 25.1L 12.21H 14.93L 110.92H 26.71L 13.4H 198.49L 11.95H 22.79L 107.25H 25.24L 15.31H 212.5L 9.62H 19.36L 10.88H 12.95L 112.5H 11.72L 16.24H 15.77L 8.43H 34.97L 3.96H 212.15L 40.78L 7.27H 36.94L 12.97H 20.45L 18.94H 15.28L 24.33H 13.48L 22.01H 19.22L 12.64H 38.5L 7.59H 225.9L 6.98H 39.31L 11.67H 45.42L 12.05H 106.47L A

This thin film consisted of 91 layers with total thickness of 4551.6 nm.

Figure 2 shows the transmittance for the two thin films.

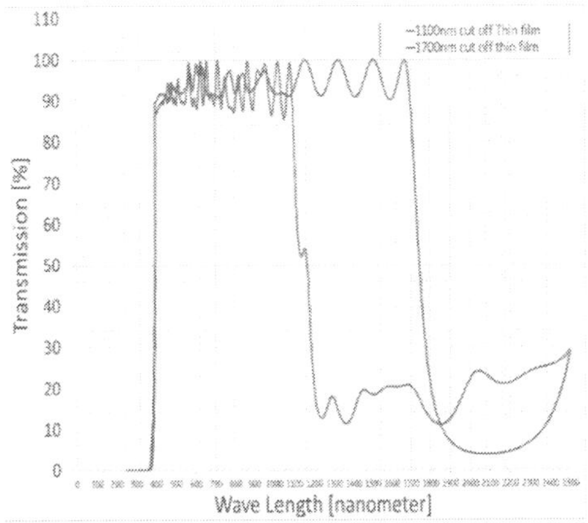

Figure 2. Thin film transmittance curve as a function of wavelength.

4 Heat Transfer module of CPV module

Heat transfer simulation for the CPV module was carried out using COMSOL Multiphysics. Figure 3 shows the geometry model developed for the calculating of heat transfer in the CPV module. The receiver consisted of homogenizer, III V solar cell, a solder, a copper electrode, insulation materials and aluminium stage which was mounted on the aluminium chassis.

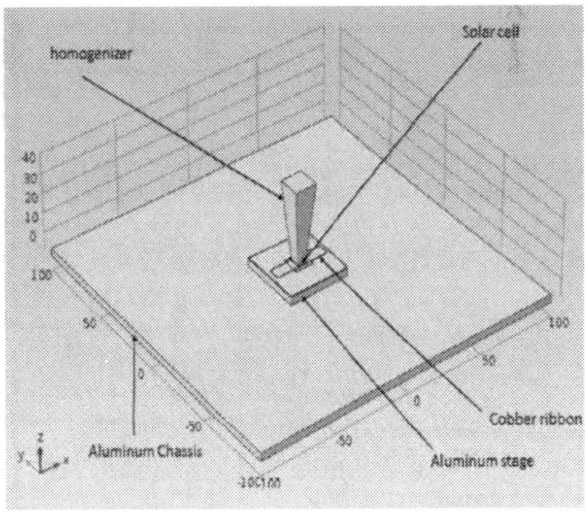

Figure 3. Simple geometry module for the heat transfer simulation.

I.Antón et al. [7] reported that the cell temperature (T_{cell}) could be related to the module temperature (temperature at the back surface of CPV module) through:

$$T_{cell}=T_{module}+R_{th\_cell\_heatsink}\times P \quad , \qquad (1)$$

where T_{module} (K) is module temperature at back surface, $R_{th\_cell\_heatsink}$ (K/W) is thermal resistance between the cell and the back chassis or heat sink core, and P (W) is the heat power, respectively. The heat Power was related to the direct normal irradiance (DNI) through:

$$P=DNI\times A_{cell}\times C\times \eta_{op}\times(1-\eta) \quad , \quad (2)$$

where A_{cell} is the solar cell area, (C) is the concentration ratio, η_{op} is the optical efficiency, and η is the electrical efficiency. A_{cell} was fixed to 5.5×5.5 mm^2. The initial temperature was 300 K.

5 Results and discussion

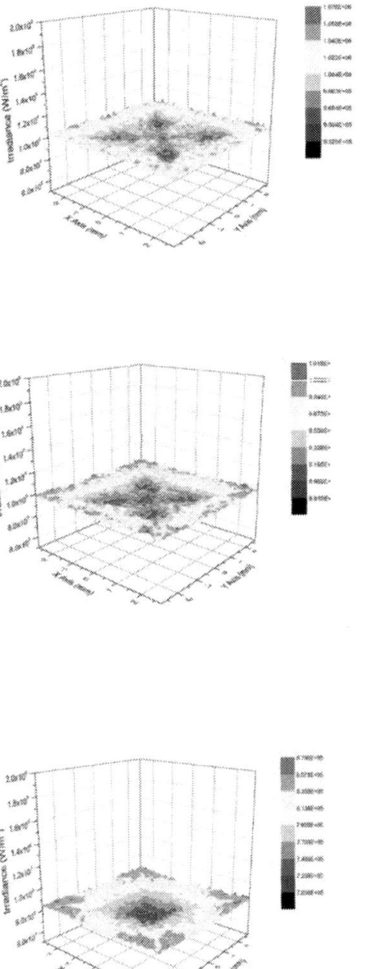

Figure 4. Calculated irradiance distribution on the solar cell.

Using ray-trace simulation we analysed the irradiance distribution on the solar cell for three cases; a)No thin film on the homogenizer face, b)1700 nm cutoff wavelength thin film, and c)1100 nm cutoff wavelength thin Film. Figure 4 shows the irradiance distribution on the solar cell for these three cases. The results showed good irradiance uniformity due to the use of homogenizer in all three cases.

Heat transfer simulation for the CPV module was carried out using COMSOL Multiphysics. The receiver consisted of homogenizer, III-V solar cell, a solder, a copper electrode, insulation materials and aluminium stage mounted on the aluminium chassis. Figure 5 shows the temperature distribution on the solar cell for each case. The results of thermal simulation for the three cases showed that the highest temperature was observed for the case of no thin film (118.7°C), while the temperature dropped to 114.5°C in the case of 1700 nm cutoff wavelength thin film with total drop of about 4°C in temperature.

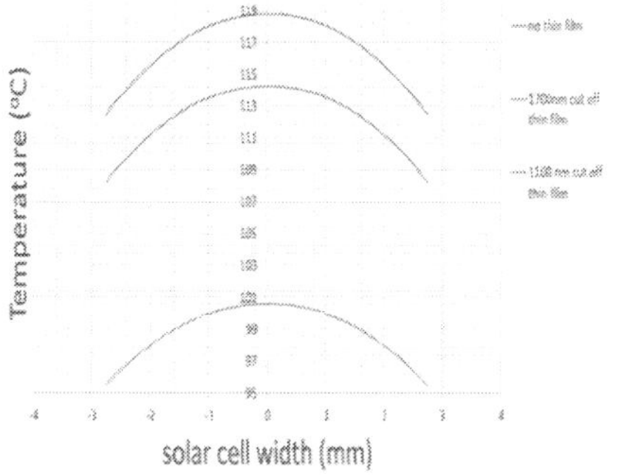

Figure 5. Temperature gradient on the solar cell surface.

The best thermal performance was achieved by the 1100 nm cutoff wavelength thin film case where the temperature decreased by 18.2°C to 100.5°C. The use of the thin film will have great effect on the solar cell performances.

6 Conclusion

Optical and thermal simulation have been conducted to an optical system with two types of multi junction solar cell and thin film that allows only the spectrum in which the solar cell has the highest response to pass through. Results showed great increase in the thermal performance of the InGaP/GaAs/InGaAs triple junction with decrease in total temperature by about 18.2oC from the no thin film case. The use of the thin film will have great effect on the solar cell performances.

References:

1. K. Araki, M. Yamaguchi, Sol. Energy Mater. Sol. Cells **75**, 467 (2003).
2. M. Bosi, C. Pelosi, Prog. Photovolt: Res. Appl. **15**, 51 (2007).
3. W. Guter, J. Schöne, S. P. Philipps, M. Steiner, G. Siefer, A. Wekkeli, E.Welser, E. Oliva, A. W. Bett, and F. Dimroth, Appl. Phys. Lett. **94**, 223504 (2009).
4. K. Sasaki, T. Agui, K. Nakaido, N. Takahashi, R. Onitsuka, and T. Takamoto, AIP Conf. Proc. **1556**, 22 (2013).
5. A.V. Tikhonravov, M.K. Trubetskov, and G.W. DeBell, Applied Optics. **35**, 5493 (1996).
6. A.V. Tikhonravov, M.K. Trubetskov, and G.W. DeBell, Applied Optics. **46**, Issue 5, 704 (2007).
7. I. Anton, M. Martinez, F. Rubio, R. Nunez, R. Herrero, C. Dominguez, M. Victoria, S. Askins, and G. Sala, AIP Conference Proceedings (8th International Conference on Concentrating Photovoltaic Systems), **1477**, 331 (2012).

Outdoor Performance Comparison of Concentrator Photovoltaic and Flat Plate Photovoltaic Systems

Yoshihide Hidaka[1,a], Kenta Imai[1],Yasuyuki Ota[1]and Kensuke Nishioka[1]

[1]*Faculty of Engineering, University of Miyazaki, 1-1 Gakuen-kibanadai-nishi, Miyazaki 889-2192, Japan*

Abstract. Output characteristics of tracking type concentrator photovoltaic (CPV) system, multi-crystalline silicon (mc-Si) PV system, CIGS PV system, and amorphous silicon (a-Si) PV system were analyzed in the data period of a year from August 2013 to July 2014.In this study, we analyzed the influence of environmental factors using average photon energy (APE) and temperature of solar cell (T_{cell}). The characteristics of 14 kW CPV system, 50 kW mc-Si PV system, 60 kW CIGS PV system, 1.35 kW a-Si PV system were evaluated and compared. As a result, the output performance of CPV was highest between the four systems at the most frequent conditions in the outdoor environment.

1 Introduction

Multi-junction solar cells have attracted increasing attention for application in concentrator photovoltaic (CPV) system owing to their very high conversion efficiency [1]. Multi-junction solar cells consisting of InGaP, InGaAs, and Ge diodes are recognized as super high efficiency cells and are used for space application. Light concentration is one of the most important factors for the development of an advanced PV system using high-efficiency solar cells. High-efficiency multi-junction cells under high light concentration have been investigated for terrestrial application [2,3].

In Japan, there are a few reports in field-test data of CPV systems. Therefore, it is important to accumulate the field-test data of CPV systems and evaluate them. PV systems which are generally used are crystalline-silicon (c-Si) PV system, CIGS PV system and amorphous silicon (a-Si) PV system.

The conversion efficiency of solar cell decreases with increasing temperature [4,5]. Characteristics of CPV are more sensitive to spectrum as compared to flat-plate PV because CPV systems use lens and multi-junction solar cells.

In this study, we analyzed the influence of environmental factors in Japanese meteorological conditions using an index of average photon energy (APE) and temperature [6]. Moreover, we compared CPV, multi-crystalline silicon (mc-Si) PV, CIGS PV, a-Si PV.

2 Experimental Procedure

Fig.1 (a) shows a 14 kW CPV system (Daido Steel Co., Ltd.) installed in the University of Miyazaki. Panel area was 7 m × 10 m. A Fresnel lens (160 mm ×160 mm) concentrates sunlight and concentrated light is irradiated onto a triple-junction solar cell (7 mm × 7 mm). A CPV module was fabricated by connecting 25 lens-cell pairs in series. The 14 kW CPV system is composed of 96 modules. Fig.1 (b) shows a 50 kW multi-crystalline Si (flat plate) PV system. The panel area of this system was 381 m^2. Fig.2 (c) shows a 60 kWCIGS (flat-plate) PV system. The panel area of this system was 563 m^2. Fig.2 (d) shows a 1.35 kWa-Si (flat-plate) PV system. The panel area of this system was 22 m^2. These 50 kW mc-Si PV system, 60 kW CIGS PV system and 1.35 kW a-Si PV system were also installed in the same site.

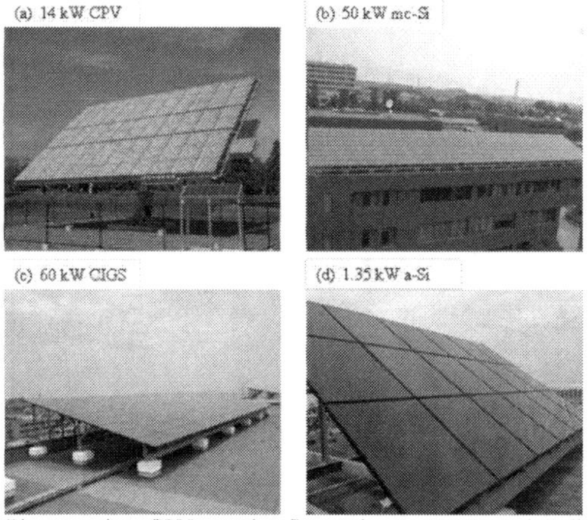

Figure 1. CPV and flat plate systems installed inUniversityofMiyazaki.

As meteorological data, we measured direct normalized irradiance (DNI), global irradiance (GI), direct solar spectra, and global solar spectra every 1

[a]hk12040@student.miyazaki-u.ac.jp

minute. DNI and GI were respectively measured by the DNI meter and GI meter. The solar spectrum was measured with the wavelength range from 350 to 1050 nm, the solar spectrum was recoded every 1 minute by a spectro-radiometer (MS700, EKO). Direct solar irradiance was recorded by a spectro-radiometer with a collimation tube (aperture angle:±2.5°).

APE is an index that indicates a spectral irradiance distribution [6]. A value of APE was calculated from measured value of spectral irradiance by dividing the integrated irradiance withthe integrated photon flux density, yielding the average energy per photon(eV):

$$APE=\frac{\int_a^b E(\lambda)d\lambda}{q\int_a^b \Phi(\lambda)d\lambda} \qquad (1)$$

where*q*is the electronic charge, *E*spectral irradiance and *Φ*the spectral photon flux density, respectively. For the limitation of our measurement system, aandbwereset to 350 and 1050 nm, respectively. For the CPV system, Fig. 2 shows a flow chart of how the contour graphs were made from APE, T_{cell}, output energy (P_{out}), and DNI, respectively. The methodology is as follows:

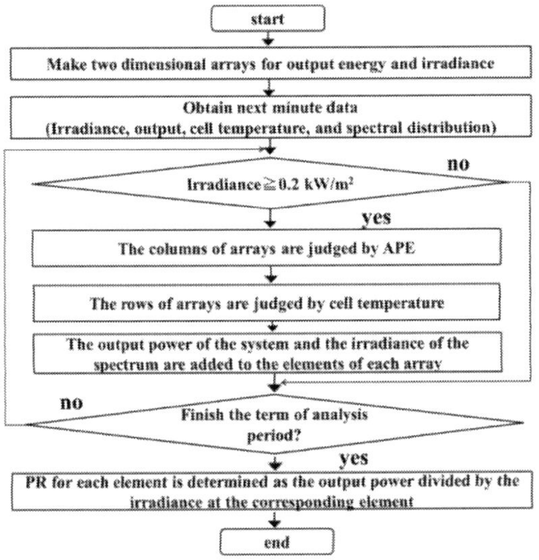

Figure 2. Flow chart of how the contour map was made.

(1) The data of T_{cell}, DNI for 1 m^2, P_{out} of the system which has nominal output power of 14kW, and the spectral irradiance distribution at the analysis period are prepared. The two-dimensional arrays for DNI and for P_{out} are prepared. The arrays for P_{out} and DNI have APE (1.80-1.99eV, 0.01 eV step) and T_{cell} (0-80°C, 5°Cstep) for column and row, respectively.

(2) The following steps are repeated for the data measured every 1 minute from August 2013 to July 2014. Here, the data with DNI higher than 0.20 kW/m^2 were used and lower performance of system at low DNI was omitted. The reason why we used irradiance higher than 0.20 kW/m^2 was the fluctuation of output voltage, which resulted in fluctuation of the maximum power point tracking (MPPT) control of the inverter.

(a) APE of spectrum is determined by the spectrum distribution.
(b) The columns of both P_{out} and DNI arrays for the spectrum are indexed by APE.
(c) The rows of both P_{out} and DNI arrays for the spectrum are indexed by T_{cell}. Then, the elements of the arrays for the spectrum are determined.
(d) P_{out} of the system and DNI of the spectrum are added to the elements of each array.

(3) Performance ratio (PR) is determined as P_{out} divided by DNI at the corresponding element. PR is given as follows:

$$PR=\frac{\frac{P_{out}(kW)}{DNI\left(kW/m^2\right)}}{\frac{P_{max}(kW)[\text{rated power of CPV system}]}{1\left(kW/m^2\right)[\text{DNI for rating}]}} \qquad (2)$$

where P_{max} is the nominal maximum output power under the standard test condition. For the mc-Si, CIGS and a-Si systems, the same method was used for making contour map. GI and solar spectrum without collimation tube were used.

3 Results and Discussion

Figure 3 shows the contour maps of PR of CPV, mc-Si PV, CIGS PV, a-Si PV systems as function of APE and T_{cell}. In the case of mc-Si, the value of PR decreases with increasing temperature owing to decrease of open-circuit voltage (V_{oc}) [7]. In the case of CPV, the decrease of PR with increasing temperature gets balanced out with the increase of performance owing to the high DNI [8].

In the case of CIGS, the value of PR decreases with increasing T_{cell} as mc-Si when APE was more than 1.90 eV. However, PR decreases with decreasing APE when APE was less than 1.88 eV. In the case of a-Si, the value of PR increases with increasing temperature owing to anneal effect [9].

Table 1 shows the most frequent conditions for the four systems and the value of PR in four systems at these conditions.

The PR of CPV at most frequent condition had the highest value.

These results indicated that the importance of the understanding of the behavior of the outdoor performance in various PVs and the accurate data of environmental conditions where various PVs are installed.

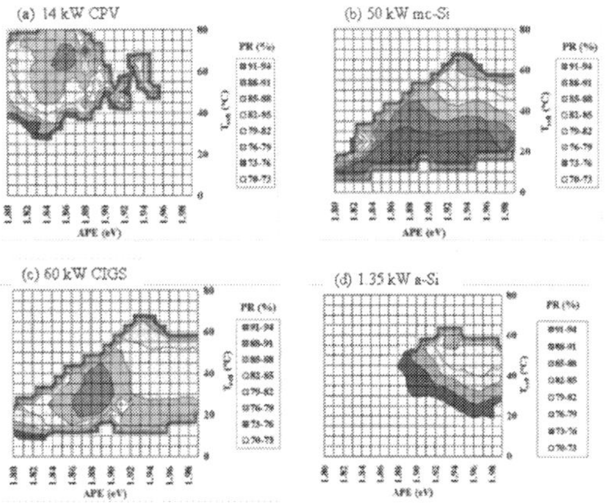

Figure 3.The outdoor performance of CPV and flat plate systems as function APE and T_{cell}

Table 1
The value of PR at the most frequent conditions in four systems

	CPV	mc-Si	CIGS	a-Si
Most frequent condition (APE(eV), T_{cell} (^{o}C))	1.830 ± 0.005, 60 ± 5	1.920 ± 0.005, 45 ± 5	1.920 ± 0.005, 45 ± 5	1.920 ± 0.005, 40 ± 5
PR (%)	86	84.5	84.3	77.3

4 Conclusion

We analyzed the influence of environmental factors in Japanese meteorological conditions using an index of average photon energy (APE) and temperature. Outdoor performance characteristics of four systems were different. The CPV and mc-Si PV systems were depended to APE and cell temperature, respectably. However, the characteristic of CIGS PV system was intermediate between CPV and mc-Si. The value of PR of CPV at the most frequent condition was higher than others. The characteristic of a-Si PV system increased owing to anneal effect.

References

1. M. Yamaguchi, Sol. Energy Mater. Sol. Cells.**75**, 261 (2003).
2. H. L. Cotal, D. R. Lillington, J. H. Ermer, R. R. King, and N. H. Karam: *Proc. 28th IEEE Photovoltaic Specialists Conf* (Anchorage, 2000).
3. A. W. Bett, F. Dimroth, G. Lange, M. Meusel, R. Beckert, M. Hein, S. V. Riesen, and U. Schubert,*Proc. 28th IEEE Photovoltaic Specialists Conf*(Anchorage, 2000).
4. S. M. Sze,*Physics of Semiconductor Devices, Wiley-Interscience Publication*(New York, 1981).
5. A.L. Fahrenbruch, R.H. Bube, *Fundamentals of Solar Cells, Academic Press Inc*(New York, 1983).
6. T. Minemoto, S. Nagae, H. Takakura, Sol. Energy.**91**, 919(2007).
7. T. Minemoto, Y. Nakada, H. Takahashi, H. Takakura, Sol. Energy.**83**, 1294(2009).
8. K. Nishioka, T. Takamoto, T. Agui, M. Kaneiwa, Y. Uraoka, Sol. Energy.**90**, 57 (2006).
9. T. Ishii, K. Otani, T. Takashima, K. Ikeda, Progress in PHOTOVOLTAICS. **22**, 949 (2014).

Reduction of Temperature in Concentrator Photovoltaic Module Using Coating with High Thermal Emissivity and Conductivity

Satoshi Nakamura[1], Kazumasa Sakamoto[1,a] and Kensuke Nishioka[1]

[1]*Faculty of EngineeringUniversity of Miyazaki, 1-1, GakuenKibanadai-nishi, Miyazaki-shi, 889-2192, Japan*

Abstract. The temperature of solar cells considerably increases under light-concentrating operations, and the conversion efficiency of solar cells decreases with increasing temperature.It is very important to reduce the cell temperature in concentrator photovoltaic modules. The thermal radiation layers with high thermal emissivity and thermal conductivity was coated on the aluminum samples which is used for the chassis of concentrator photovoltaic and the effect was evaluated under the conditions with wind. The temperature of sample with coating showed lower temperature than that without coating. In the condition with wind, the coated sample with high thermal emissivity and high thermal conductivity showed the lowest temperature due to the effect of thermal radiation and thermal conduction.

1Introduction

In recently year, multi-junction solar cells have attracted increasing attention for application in concentrator photovoltaic (CPV) system owing to their very high conversion efficiency [1-3]. Multi-junction solar cells consisting of InGaP, InGaAs, and Ge diodes are recognized as super high efficiency cells and are used for space application. Light concentration is one of the most important factors for the development of an advanced PV system using high-efficiency solar cells. High-efficiency multi-junction cells under high light concentration have been investigated for terrestrial application [4-7].The temperature of solar cells considerably increases under light-concentrating operations, and the conversion efficiency of solar cells decreases with increasing temperature [8-10]. It is therefore very important to reduce the cell temperature in CPV modules. In this study, thermal radiation layers with high thermal emissivity and thermal conductivity was coated on the aluminum samples which is used for the chassis of CPV and the effect was evaluated under the conditions with wind.

2 Experimental

Fig.1 shows the samples used in this study. Aluminum samples(60 mm width x 95 mm lengthx 3 mm thickness) wereprepared. In order to insert a thermalsensor, a hole of 1.1mm in diameter with 10mm in depth was made at thecenter of cross-section. In order to detect the sample temperature, Pt100 was inserted in the hole. The heat conductive grease was used in the hole to detect the exact temperature. Three aluminum plates were prepared; "high emissivity type" with high thermal emissivity material, "high emissivity and conductivity type" with high thermal emissivity and high thermal conductivity material and "reference" without coating.The thermal radiation layer [Pelcool (R), PELNOX Ltd.] was coated on the aluminumwhich is used as the material of the chassis of CPV module by a spray coating method. The thickness of the layer was 30 μm. The layer consisted of acrylate resin and inorganic fillers. Heater was fixed with double faced tape on each sample.

Fig.2 shows a schematic diagram of sample setup. Three samples were connected with parallel circuit and bias was applied using a biporal power supply with constant voltage operation. The bias voltage was set at 80 V and the current value was 0.97 A. The temperature was measured every second and accumulated by a data logger.

Fig.3 shows a schematic diagram of wind tunnel used in this study. Total of 99 pieces AC fans were installed in the wind tunnel, 11 fans in depth and, 9 fans in widths, and they make air flow. Each fan is set directto the servomotor. It is dependently suppressed from spinning by the signalgenerated from a computer every 0.02 seconds. Theturbulence strength is under 2.0%. It proves the high performance of thewind tunnel [11].The size of the wind tunnel was1,800 mm in height and 2,540 mm in width and the distance from blast fan to the samples was 5.6m. Wind speed was measured with pitot tube that was set at the 1.23m front positon of the samples.

[a]ha11012@student.miyazaki-u.ac.jp

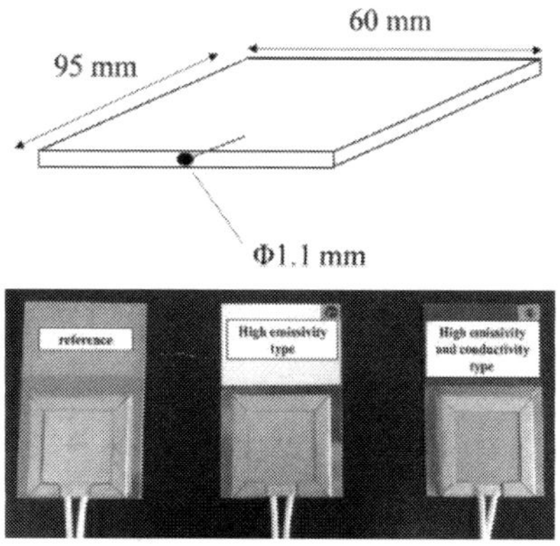

Figure 1. Samples used in this study.

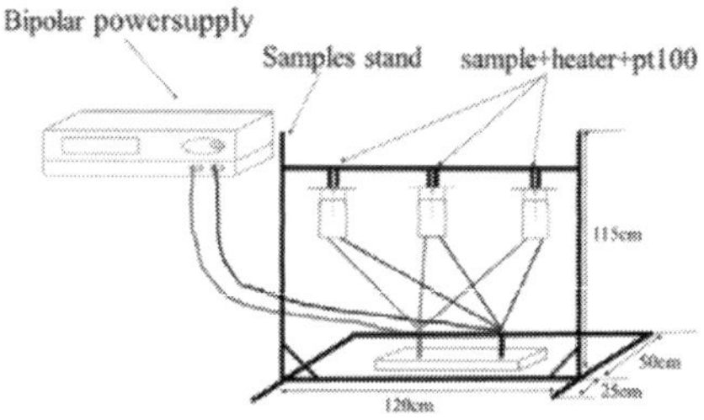

Figure 2. Schematic diagram of sample setup.

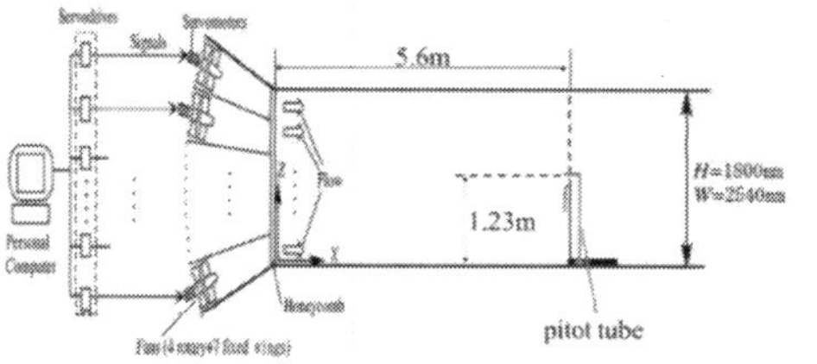

Figure 3. Schematic diagram of wind tunnel.

3Results and Discussion

Fig. 4 shows the temperature of samples for each wind speed. The temperature with coating showed lower temperature than that without coating. The thermal radiation coatings worked well as heat reduction coating.

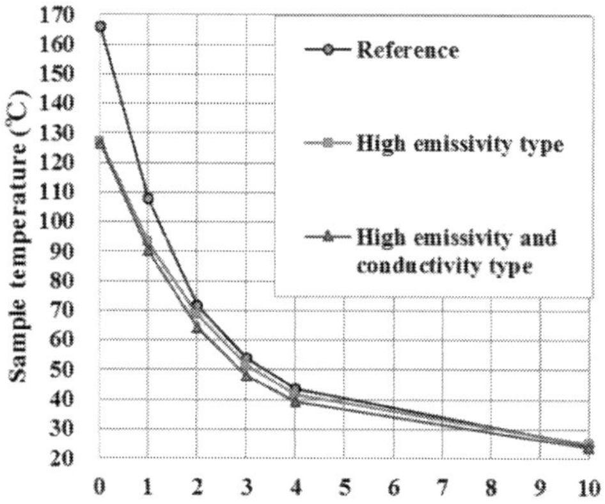

Figure 4. Temperature of samplesfor each wind speed.

When we define a layer of sample's surface [area: A_1 (m^2), emissivity: ε_1] and an environment [area: infinite], the heat radiation from the sample to the environment Q_{12} (W) is given by

$$Q_{12}=A_1\varepsilon_1\sigma(T_1^4-T_2^4) \quad (1)$$

where σ, T_1, and T_2 are the Stefan–Boltzmann constant (5.67×10^{-8} W/m^2·K^4), the absolute temperature (K) of the sample, and the absolute temperature (K) of the environment, respectively. It was found that the heat radiation increased with increasing emissivity, and the high emissivity of the thermal radiation layer enhanced the heat radiation from the sample to the environment.For all samples, the temperature decreased with increasing wind speed. It is found that wind is important to decrease the sample temperature.

Fig.5 shows the difference in temperature between "high emissivity and conductivity type" and "high emissivity type" for each wind speed ("difference in temperature" = "temperature of high emissivity and conductivity type"–"high emissivity type"). In the non-wind condition, there is no significantdifference between them. However, with the wind conditions, the difference in temperature became larger with increasing wind speed. When we define a layer of sample's surface [area: A_1 (m^2),thickness: L_1 (m), conductivity: λ_1], the thermal conduction of inside the layer$\dot{Q}_{12}$ (W) is given by

$$\dot{Q}_{12}=A_1\lambda_1\frac{T_1-T_2}{L_1} \quad (2)$$

An increase in thermal conductivity accelerates the heat transfer from the alminum to enironment.Therefore, the temperautre of the sample with high emissivity and conductivity type was lower than that with high emissivity type. However, in the wind condition over 3 m/s, the difference in temperature became smaller. It is considered that the heat reduction effect of wind is preferable in the wind condition of over 3 m/s.

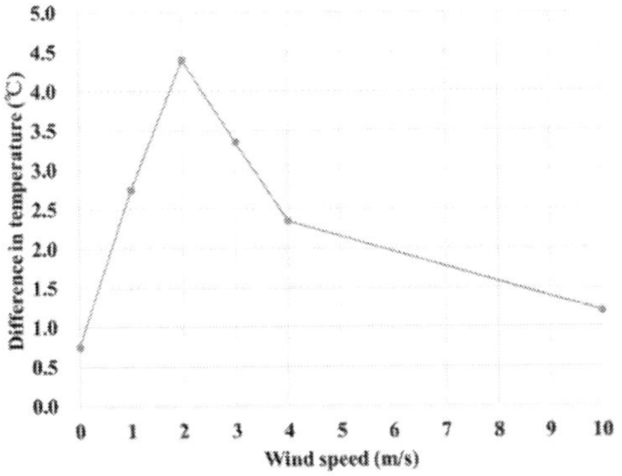

Figure 5. Difference in temperature between high emissivity and conductivity type and high emissivity type for each wind speed (difference in temperature = temperature of high emissivity and conductivity type - high emissivity type).

4 Summary

The thermal radiation layers with high thermal emissivity and thermal conductivity was coated on the aluminum samples which is used for the chassis of CPV and the effect was evaluated under the conditions with wind. The temperature with coating showed lower temperature than that without coating. For all samples, the temperature decreased with increasing wind speed. It is found that wind is important to decrease the sample temperature. The temperautre of the sample with high emissivity and conductivity type was lower than that with high emissivity type.

References

1. M. Yamaguchi,Sol. Energy Mater. Sol. Cells. **75**,261(2003)
2. M. Yamaguchi, T. Takamoto, K. Araki,Sol. Energy Mater. Sol. Cells.**90,**57(2006)
3. T. Takamoto, M. Kaneiwa, M. Imaizumi, M. Yamaguchi,Prog. Photovoltaics Res. Appl. **13**,495(2005)
4. K. Araki, M. Kondo, H. Uozumi, N.J. Ekins-Daukes, T. Egami, M. Hiramatsu, Y. Miyazaki, M.Yamaguchi,Sol. Energy Mater. Sol. Cells.**90**,3320(2006)
5. J. Jaus, A.W. Bett, H. Reinecke, E.R. Weber,Prog. Photovoltaics Res.Appl. **19**,580(2011)
6. K. Nishioka, T. Takamoto, T. Agui, M. Kaneiwa, Y. Uraoka, T. Fuyuki,Sol. Energy Mater. Sol. Cells.**90**,57(2006)
7. K Araki, T. Yano, Y. Kuroda,Optics Express.**18** A53(2010)
8. K. Nishioka, T. Hatayama, Y. Uraoka, T. Fuyuki, R. Hagihara, M. Watanabe,Sol. Energy Mater. Sol. Cells.**665**,75 (2003)
9. K. Nishioka , N. Sakitani, K. Kurobe, Y. Yamamoto, Y. Ishikawa, Y. Uraoka, T. Fuyuki,Jpn. J.Appl. Phys. **42**,7175(2003)
10. K. Nishioka, T. Takamoto, T. Agui, M. Kaneiwa, Y. Uraoka, T. Fuyuki,Sol. Energy Mater. Sol. Cells.**85** 429 (2005)
11. S. Ozono, A. Nishi, H. Miyagi,, J. of Wind Engineering & Industrial Aerodynamics.**94** 225(2006)

Photochemical Synthesis of Au@Pd Core-Shell Nanoparticles for Methanol Oxidation Reaction: the Promotional Effect of the Au Core

Yingnan Dong[1,2,a] Xikun Yang[3] Zhengfu Zhang[3] Shouan Dong[2] and Shanshan Li[1]

[1]*Shenyang Institute of Engineering, Shenyang 110136, China*
[2]*Kunming Institute of Precious Metals, Kunming 650106, China*
[3]*Kunming University of Science and Technology, Kunming 650091, China*

Abstract. A novel method for synthesizing Au@Pd core-shell nanoparticles was proposed based on photochemistry. By irradiating the mixture of Au (III) and Pd (II) ions using ultraviolet light, the Au@Pd core-shell nanoparticles were prepared. The size of the nanoparticles and the thickness of the Pd shell could be efficiently adjusted by changing the molar ratio of Au (III) to Pd (II) ion. In this way, nanoparticles with diameter in the range of 5.6~4.6 nm were obtained. The core-shell structure of the synthesized nanoparticles was showed by the characterization using UV-Vis, TEM/HR-TEM and XPS. The paper investigated the electrocatalysis performance of Au@Pd nanoparticles in the methanol catalytic oxidation reaction, as well as the electron donating effect of Au core to Pd shell and the promotion of this effect on the catalytic activity of Pd shell. The experimental results provided reference for the development of non-platinum catalysts of low-temperature fuel cell anode.

1 Introduction

Low-temperature fuel cells, including proton exchange membrane fuel cell (PEMFC), direct methanol fuel cell (DMFC), etc, are a kind of green energy technology which transfers the chemical energy of the fuel directly to electric energy without combustion. At present, the cells are of great significance in solving the energy shortage and environmental pollution worldwide and regarded as one of the most important energies in 21th century. In the cells, the electrode catalyst, particularly anode catalyst generally uses platinum (Pt) -based catalysts. In recent years, owing to the promising application prospect in electronics and optics, and the catalytic effect, the bimetalic nanoparticles have been considered to be widely used in electonics and optics , and the catalysts, and have drawn a lot of attentions.[1,2]. Among which Au@Pd nanoparticles, which are used as catalysts in low-temperature fuel cells, have been a research focus. Jim Yang Lee et al[3]conducted HCOOH oxidation reaction by applying Au@Pd core-shell nanoparticles and found that the nanoparticles showed a high catalytic activity in perchloric acid medium; the small size of Au core possibly promoted the catalytic activity.

Regarding the preparation of Au@Pd core-shell composite nanoparticles, many methods have been studied intensively, including chemical reduction method[4], sonochemical method[5] and microwave method[6]. However, the synthesis method which can precisely control the size of the composite nanoparticels and the thickness of the shell, with simplified, economic, and green process, has seldom been reported. The paper presented a novel method for synthesizing Au@Pd core-shell nanoparticles based on photochemistry. The mixture of Au (III) and Pd (II) ions was irradiated by ultraviolet light of 300 nm in the PEG (polyethylene-glycol)-acetone solution system (co-reduction). It was showed from characterization using UV-Vis, TEM, HR-TEM and XPS that the core-shell nanoparticles with uniform size and monodispersity were obtained. Based on the electrochemical workstation, the paper investigated the electrocatalysis performance of Au@Pd nanoparticles for the methanol catalytic oxidation reaction (MOR), as well as the electron donating effect of Au core on the dissimilar metal (Pd) shell and the promotion of this effect on the catalytic activity of Pd shell.

2 Experimental methods

2.1 Preparation of products

Certain amounts of PEG-400, acetone and different molar ratios of $HAuCl_4$ and $PdCl_2$ (the molar ratios of Au to Pd elements are 4:1, 1:1 and 1:4, respectively, and the total molar amount of noble metal is 1.22×10-5 mol). Then water was used to get the constant volume. After being mixed sufficiently, the solution was transferred into a quartz container and then exposed 4 cm away under ultraviolet light. UV-Vis spectra were recorded as time passed by until the spectral curve was the same as the previous one, which indicated that the reduction finished.

[a] Corresponding author: dyn4870@163.com

The single Au and Pd nanoparticles were prepared similarly. All the reagents were analytical pure and the water was ultrapure.

2.2 Characterization

Images of the prepared nanoparticles were taken by transmission electron microscopy (TEM). The sample was prepared by dropping several drips of colloidal solution on carbon support film copper grid and dried at room temperature. Then the images were taken on H-800 TEM which operated at a voltage of 200 kV. By adopting lacey support film grid, HR-TEM detection was conducted, and the images were taken on Tecnai G^2 F20 field emission TEM operated at a voltage of 200 kV. Meanwhile, XPS was performed on the sample as well. The obtained nanoparticles were loaded on XC-72 carbon black for electro-chemical measurements based on the method reported in literature[7]. The electro-chemical measurement was conducted at 30℃, and three-electrode system was adopted with Pt as the counter electrode and SCE (Saturated calomel electrode) as reference electrode. The dissolved oxygen in the electrolyte was 0.5 mol•L^{-1} KOH + 2.0 mol•L^{-1} CH_3OH. Before the measurement, high pure N_2 was bubbled into the electrolyte to remove the dissolved oxygen and then CV curves were recorded in the N_2 atmosphere. The measuring conditions are as follows: potential is in the range -0.6~0.8 V, potential scan rate is 50 mV•s^{-1}, and I-t curves are tested at electric potential of -0.3 V.

3 Results and discussion

3.1 Characterization of Au@Pd nanoparticles

In the selected solution system, the surface plasmon resonance (SPR) absorption spectra of obtained Au@Pd colloidal solution were shown in Figure 1a, and the absorption spectra of Au and Pd colloidal solutions and their mixture are displayed in Figure 1b. It is clearly demonstrated that the transition feature from Au to Pd of the synthesized Au@Pd bimetallic colloidal particles; while the latter one shows distinct superposition property. This indicated that in the co-reduction of Au (III) and Pd (II) ions based on photochemistry, owing to $AuCl^{4-}$ presents higher reduction potential ($AuCl^{4-}$/Au E^0=+1.0 V; $PdCl_4^{2-}$/Pd, E^0=+0.6 V), Au was photoreduced as the seed in priority and autocatalyzed to nanoparticles; afterward, the Pd (II) ions in the solution were catalyzed on the surface of Au nanoparticles, therefore forming the Au@Pd core-shell composite nano-structure. Considering the Au@Pd nanoparticles, as the molar ratio of Au to Pd reduces, the Au seeds are coated by Pd atoms gradually. When Au:Pd=1:1, the characteristic absorption band of Au does not completely disappear in the SRP absorption spectra of composite particles. It indicates that the Pd shell is still thin or the Pd does not completely coat on Au seed. This was basically consistent with the results in literature[4].

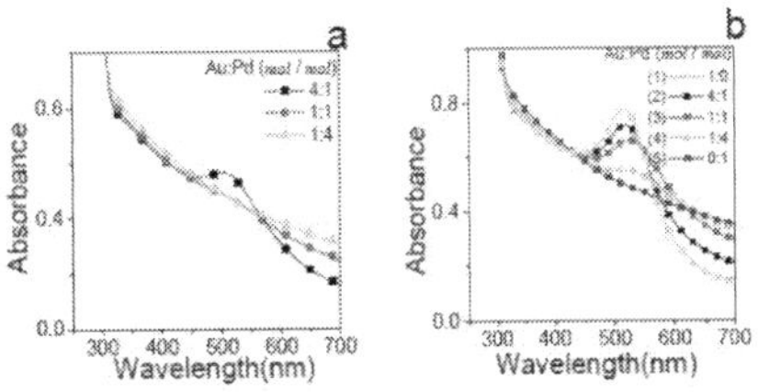

Figure 1. UV-Vis spectra of Au@Pd (a) and Au, Pd and their mixture (b) colloidal solution.

The structure of Au@Pd composite nanoparticles was analyzed using X-ray photoelectron spectroscopy (XPS). The results are listed in Table 1. It is observed that the molar ratio of Au to Pd in the surface structure of the nanoparticles is much less than that in the solution in the preparation. This states that there is higher Pd atomic fraction in the surface structure of Au@Pd nanoparticles and further proves the core-shell nano structure of the as-prepared Au@Pd. The results are in line with literature[6] and the above characterization using UV-Vis absorption spectra.

Table 1. XPS analysis of Au@Pd nanoparticles

and Pd $3d_{5/2}$ binding energy.

Au:Pd (molar ratio)		4:1	1:1	1:4
Atom ratio on the surface	Au%	67.4	32.1	6.6
	Pd%	32.6	67.9	93.4
Pd $3d_{5/2}$ Binding Energy (eV, E_0= 335.10 eV[8])		333.84	334.59	334.64
Δ E (Binding Energy, eV;)		-1.26	-0.51	-0.46

In the linear fitting of spectral lines of the XPS energy spectra of the samples, it is observed that there are two spectral peaks in Pd 3d electronic spectrogram (Figure 2). Compared with the energy spectra[8] of standard palladium, the peak near 334 eV is simple substance Pd (Pd $3d_{5/2}$, 335.1 eV); while the one around 336 eV is oxidized Pd (PdO $3d_{5/2}$, 336.3 eV). This illustrates that the Pd atoms on the surface of Au@Pd composite particles are prone to be oxidized. Furthermore, the binding energy of Pd in Au@Pd particles decreases with the decrease of molar ratio of Au to Pd (Table 1), indicating that the Au core supplies electrons for the Pd on the external surface, and the electron donating intensity is associated with the ratio of core to shell.

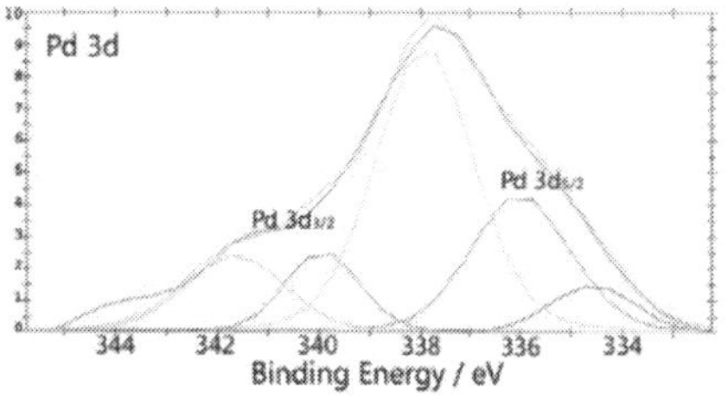

Figure 2. XPS spectra of Pd 3d of Au@Pd particles (Au:Pd = 1:1).

In the aforementioned molar ratio of Au to Pd condition, the images of synthesized Au@Pd particles taken by the transmission electron microscopy (TEM) are illustrated in Figure 3. The Au@Pd particles with uniform size and good dispersion are observed from the figures, and the average diameters are 5.6, 5.0 and 4.6 nm, respectively. The inset in Figure 3b is the HR-TEM image. Lattice fringes are clearly observed from the spherical Au@Pd particles of 5 nm, which represents that the composite particles are nano-crystalline. However, owing to the similar metallic lattice constants of Au and Pd, these two atoms are prone to mutually permeate to form solid solution. As a result, the clear metallic phase interfaces are hardly observed even from HR-TEM.

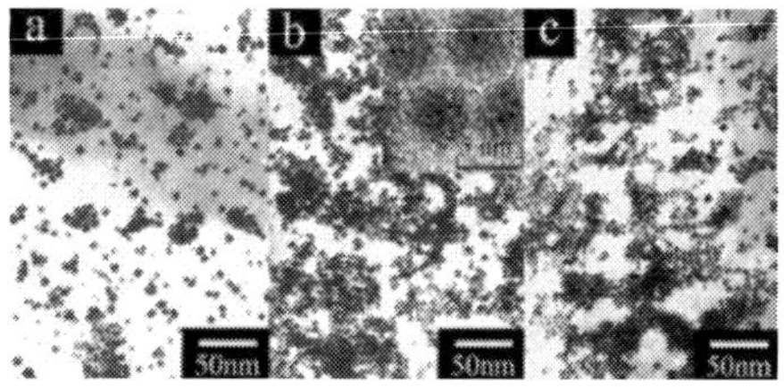

Figure 3. TEM images of Au@Pd nanoparticles: Au:Pd=4:1 (a), 1:1 (b) and 1:4 (c).

3.2 MOR activity of Au@Pd core-shell nanoparticles

Figures 4a and 4b demonstrate the cyclic voltammograms (CV) curves and time-current (I-t) curves for MOR of prepared Au@Pd/C with different molar rations of Au to Pd, single Pd/C and Pt/C catalysts in alkaline solution, respectively. Thereinto, the Pt/C was purchased from Johnson Matthey Company. The CV curves in Figure 4a illustrate that in the positive scan of all the electrodes formed by the catalyst samples, methanol is catalyzed and oxidized on the surface of electrode at ~-0.15 V, therefore generating oxidation peak currents. This indicates that all the samples present catalytic activity for MOR. Based on the calculation according to the data in Figure 4a, it is seen that the activity of pure palladium sample is apparently lower than that of platinum sample; the peak current of Pd/C merely accounts for 40% of that of Pt/C (64.4 $mA{\bullet}mg^{-1}{}_{Pd}$ and 161.2 $mA{\bullet}mg^{-1}{}_{Pt}$, respectively). While regarding Au@Pd sample, as the molar ratio of Au to Pd increases from 4:1 to 1:4, the catalytic activity of Pd increases significantly and the peak currents are 68.5, 133.0 and 181.4 $mA{\bullet}mg^{-1}{}_{Pd}$, respectively. Of which, when Au:Pd=4:1, the activity of Pd is 12.5% higher than that of Pt. This explains that Au core promotes the electrocatalytic activity of Pd shell. Furthermore, it is observed from the comparison of XPS characterization data that the promotion trend of Au core for the electrocatalytic activity of Pd shell is the same with that of Au core supplying electrons for the Pd shell. Based on the analysis and owing to Au is the last element on the 5d complete orbit, it is speculated that Au is affected by the relativistic effect most greatly[9]. Moreover, the electron contraction for 6s increases the electron cloud density and decreases the barriers of the hybridization between Au and Pd atoms. This is beneficial for the generation of large conjugated system by the interaction of Au and Pd atoms. Meanwhile, owing to the electronegativity (X=2.4) of Au is larger than that of Pd (X=2.2), part of 5d electrons of Au flows into the 4d orbit of Pd and causes part of negative charge on Pd. In this way, the shielding effect on the inner shell electrons increases, therefore reducing the electron binding energy on the inner orbit. As observed, there is a displacement of binding energy to low-end on the XPS spectra. The supplement of electrons of Au core to Pd shell efficiently improves the d-electron orbit cavity of Pd shell, which is the key reason of the outstanding catalytic activity exhibited by Au@Pd core-shell nanoparticles3. Additionally, the core-shell structure increases the specific surface area of Pd shell, which is possibly another reason for the improvement of Pd catalytic activity. The I-t curves of Pt/C and Au@Pd/C (Au:Pd=4:1) are illustrated in Figure 4b. It is observed that the Au@Pd core-shell nanoparticles exhibit a stable electrocatalysis.

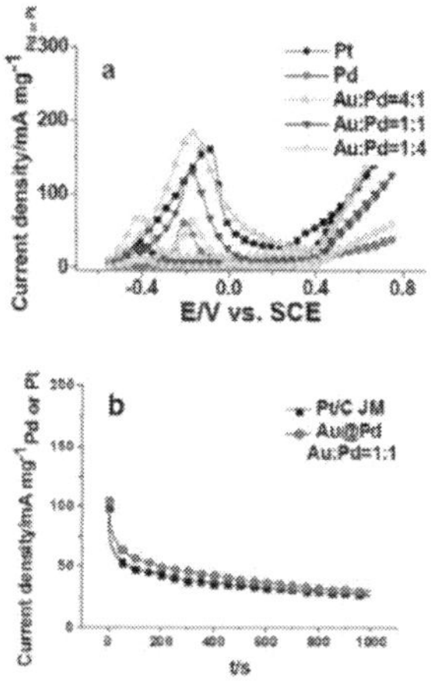

Figure 4. CV (a) and I-t (b) of Au@Pd/C in 0.5 $mol{\cdot}L^{-1}$ KOH + 2.0 $mol{\cdot}L^{-1}$ CH_3OH solution.

4 Conclusions

In the PEG-acetone system, the paper synthesized the Au@Pd core-shell nanoparticles with uniform size and monodispersion by reducing the mixture of Au (III) and Pd (II) based on photochemistry. The characterization using UV-Vis absorption spectra, TEM/HR-TEM and XPS showed that the core-shell structure nanoparticles were obtained and the Au core supplied electrons for Pd shell. The electronchemical analysis indicated that the activity of Pd for MOR increases with the increase of molar ratio of Au to Pd. Furthermore, the electron donating of Au core for Pd shell was investigated, as well as its promotion for the catalytic activity of Pd shell.

Acknowledgements

The work reported in this paper has received financial support from National Natural Science Foundation of China (No. 51164017 and 20863003).

References

1. D. Stolten, B. Emonts. Platinum Metals Rev., **66**, 57 (2013)
2. Y. N. Dong, S. S. Li, L. Cong. Yunnan Chemical Technology, **21**, 39 (2012)
3. W. L. Zhou, J. Y. Lee. Electrochem. Commun., **1725**, 9 (2007)
4. J. f. Li, J. W. Hu, B. Ren. Acta. Phys-Chim. Sin., **825**, 21 (2005)
5. C. Kan, W. Cai, C. Li, L. Zhang. J. Phys. D, **1609**, 36 (2003)
6. R. Harpeness, A. Gedanken. Langmuir, **3431**, 20 (2004)
7. J. H. Zeng, J. Yang, J. Y. Lee. J. Phys. Chem. B, **24606**, 110 (2006)
8. F. M. John. *Handbook of X-Ray Photoelectron Spectroscopy* (Perkin-Elmer Physical Electronics DiVision, 1992)
9. Y. N. Dong, S. A. Dong, C. Tang. Rare Met Mater and Eng, **789**, 42 (2013)

AUTHOR INDEX